SpringerBriefs in Applied Sciences and Technology

Nanoscience and Nanotechnology

Series Editors

Hilmi Volkan Demir, Nanyang Technological University, Singapore, Singapore

Alexander Govorov, Clippinger Laboratory 251B, Ohio Univ, Dept of Phys & Astro, Athens, OH, USA

Indexed by SCOPUS Nanoscience and nanotechnology offer means to assemble and study superstructures, composed of nanocomponents such as nanocrystals and biomolecules, exhibiting interesting unique properties. Also, nanoscience and nanotechnology enable ways to make and explore design-based artificial structures that do not exist in nature such as metamaterials and metasurfaces. Furthermore, nanoscience and nanotechnology allow us to make and understand tightly confined quasi-zero-dimensional to two-dimensional quantum structures such as nanopalettes and graphene with unique electronic structures. For example, today by using a biomolecular linker, one can assemble crystalline nanoparticles and nanowires into complex surfaces or composite structures with new electronic and optical properties. The unique properties of these superstructures result from the chemical composition and physical arrangement of such nanocomponents (e.g., semiconductor nanocrystals, metal nanoparticles, and biomolecules). Interactions between these elements (donor and acceptor) may further enhance such properties of the resulting hybrid superstructures. One of the important mechanisms is excitonics (enabled through energy transfer of exciton-exciton coupling) and another one is plasmonics (enabled by plasmon-exciton coupling). Also, in such nanoengineered structures, the light-material interactions at the nanoscale can be modified and enhanced, giving rise to nanophotonic effects.

These emerging topics of energy transfer, plasmonics, metastructuring and the like have now reached a level of wide-scale use and popularity that they are no longer the topics of a specialist, but now span the interests of all "end-users" of the new findings in these topics including those parties in biology, medicine, materials science and engineerings. Many technical books and reports have been published on individual topics in the specialized fields, and the existing literature have been typically written in a specialized manner for those in the field of interest (e.g., for only the physicists, only the chemists, etc.). However, currently there is no brief series available, which covers these topics in a way uniting all fields of interest including physics, chemistry, material science, biology, medicine, engineering, and the others.

The proposed new series in "Nanoscience and Nanotechnology" uniquely supports this cross-sectional platform spanning all of these fields. The proposed briefs series is intended to target a diverse readership and to serve as an important reference for both the specialized and general audience. This is not possible to achieve under the series of an engineering field (for example, electrical engineering) or under the series of a technical field (for example, physics and applied physics), which would have been very intimidating for biologists, medical doctors, materials scientists, etc.

The Briefs in NANOSCIENCE AND NANOTECHNOLOGY thus offers a great potential by itself, which will be interesting both for the specialists and the non-specialists.

More information about this subseries at http://www.springer.com/series/11713

Heike C. Herper · Barbara Brena ·
Carla Puglia · Sumanta Bhandary ·
Heiko Wende · Olle Eriksson · Biplab Sanyal

Molecular Nanomagnets

Fundamental Understanding

 Springer

Heike C. Herper
Department of Physics and Astronomy
Uppsala University
Uppsala, Sweden

Barbara Brena
Department of Physics and Astronomy
Uppsala University
Uppsala, Sweden

Carla Puglia
Department of Physics and Astronomy
Uppsala University
Uppsala, Sweden

Sumanta Bhandary
Centre de Physique Theorique (CPHT)
Ecole Polytechnique
Palaiseau, Paris, France

Heiko Wende
Fakultät für Physik
Universität Duisburg-Essen
Duisburg, Nordrhein-Westfalen, Germany

Olle Eriksson
Department of Physics and Astronomy
Uppsala University
Uppsala, Sweden

Biplab Sanyal
Department of Physics and Astronomy
Uppsala University
Uppsala, Sweden

ISSN 2191-530X   ISSN 2191-5318   (electronic)
SpringerBriefs in Applied Sciences and Technology
ISSN 2196-1670   ISSN 2196-1689   (electronic)
Nanoscience and Nanotechnology
ISBN 978-981-15-3718-9   ISBN 978-981-15-3719-6   (eBook)
https://doi.org/10.1007/978-981-15-3719-6

This Springer imprint is published by the registered company Springer Nature Singapore Pte Ltd.
The registered company address is: 152 Beach Road, #21-01/04 Gateway East, Singapore 189721, Singapore

Foreword

This book is a monograph covering the electronic states and magnetic properties of isolated and deposited transition-metal phthalocyanines. Phthalocyanine (H_2Pc) is an organic molecule of the formula $(C_8H_4N_2)4H_2$ and its metal complexes $M^{2+}Pc^{2-}$ are well known for many years to be useful compounds for the applications to dyes, pigments, and catalysts. Recently, transition-metal Pc complexes have attracted much attention because of novel magnetic properties and their possible tunability by changing ligand, metal center, substrate, and so on for spintronics applications. For studying the electronic and magnetic structure of the Pc compounds and complexes, several modern experimental techniques such as X-ray absorption spectroscopy (XAS), photoelectron spectroscopy (PES), and scanning tunneling microscopy (STM) are briefly introduced. XAS is an element, orbital, and symmetry specific probing technique by tuning the photon energy and polarization. Especially, magnetic circular dichroism at transition-metal $L_{2,3}$ absorption edges together with the so-called orbital and spin sum rules is a powerful tool to extract the orbital and spin magnetic moments separately from the spectra. Transition-metal XAS spectra at K edge are very sensitive to its local geometry and atomic valence. PES is usually used for investigating the electronic valence states, while the chemical shift of core electron states contains the information of environmental changes around the atom. STM is a real-space probe to look at surface on a scale of Å. To analyze the experimental data obtained above, it is now indispensable to employ first-principles density-functional-theory (DFT) calculations. The standard DFT calculations within the local density approximation (LDA) or its semi-local version are generally sufficient for determining the stable structure and vibrational properties for many materials. However, to investigate the so-called correlated electron states often appearing in transition-metal organic complexes, one must go beyond the LDA level particularly for understanding competing spin states and associated spin-crossover (spin-transition) phenomena coupled with geometry and atomic valence. Several approaches are introduced for describing magnetic molecules in an isolated form and on substrate, and electron correlation is discussed in detail. In addition, for the readers with interest of applications, some related molecular devices are explained. In conclusion, clear and concise introduction of modern experimental and theoretical/computational techniques for studying the electronic and magnetic properties of transition-metal Pc complexes as well as the underlying physics of electron correlation involved are given in this book. These contents might be worthwhile and quite valuable for graduate students and novice as the first step of the experimental and theoretical/computational approaches to the fundamentals of transition-metal Pc complexes and related materials, in wide fields of science and engineering such as condensed matter physics, applied physics, organic chemistry, molecular science, and surface science.

April 2018   Tamio Oguchi
                                                    Osaka University, Osaka, Japan

Preface

Molecular spintronics is an upcoming field of research with a huge potential towards applications in future nanotechnology. It has been realized so far that the development of spintronic devices is far from being complete as it faces many challenges partially from the point of technical realizations but also on a more basic level due to an incomplete understanding of the fundamental processes in the atomic scale. Molecules in gas phase, in this book mostly transition metal phthalocyanines, are studied for decades and their properties are now quite well understood, while the molecules deposited on substrates have attracted huge interest more recently as building blocks for spintronic devices. The so-called *spinterface* formed at the interface between an organic molecule and an inorganic substrate has become a key issue to be understood in the electronic and atomic level for realizing spin transport through the magnetic molecule. This has led to a large number of publications. However, despite the large amount of data, there is no unique understanding e.g. of the coupling mechanisms between molecule and substrate.

In this book, we provide an overview on the theoretical and experimental findings for deposited transition metal phthalocyanines and put this in context with the methods used to achieve the data. This is accompanied by a brief description of the most relevant experimental techniques and theoretical methods in this field. The book discusses the interplay between different phthlaocyanine molecules with various substrates such as simple metal surfaces, magnetic substrates and complex reconstructed surfaces. Intricate properties such as spin dipole moment and many body Coulomb correlation induced spin transition have been discussed.

This book has been designed not only for the experts in the field but is also a valuable overview for Ph.D. students to learn about the capabilities and limitations of the theoretical as well as experimental tools used for molecules in gas phase and deposited on substrates.

Uppsala, Sweden
June 2019

Heike C. Herper
Biplab Sanyal

Acknowledgements

We acknowledge financial support from Swedish Research Council, STINT and Carl Tryggers Stiftelse. Also, Swedish National Infrastructure for Computing is acknowledged for allocations of high performance supercomputing time.

Contents

1 Introduction 1
References 3

2 Experimental Techniques 5
2.1 X-ray Absorption Spectroscopy (XAS) 5
2.1.1 Coupling of FePc to Co Film 7
2.2 Photoelectron Spectroscopy (PES) 11
2.3 Adsorbate Core Level Shift: Final State Effects 13
2.4 Scanning Tunneling Microscopy (STM) 14
References 16

3 Theoretical Methods 19
3.1 Density Functional Theory 19
3.2 Hybrid Functionals 19
3.3 Electron Correlation 20
3.4 van der Waals Interaction 21
3.5 Magneto-Crystalline Anisotropy 22
3.6 Spin Dipole Moment 23
References 24

4 Electronic Structure of Isolated Molecules 25
4.1 Phthalocyanines with Transition Metal Center 25
4.2 MnPc 27
4.3 FePc 29
4.4 CoPc 31
4.5 NiPc 31
4.6 CuPc 32
References 33

5 Electron Correlation and Spin Transition   35
   5.1 Electron Correlation: DFT++   35
   5.2 Dynamical Hybridization Function   37
   5.3 The Spin Transition   38
   5.4 Kondo Effect and Hybrdization Function   41
   References   42

6 Interaction with Substrates   45
   6.1 Metal Substrates   45
      6.1.1 Structural Aspects   46
      6.1.2 Magnetic Coupling   52
      6.1.3 Spin Moments and the Spin Dipole Contribution   57
   6.2 Nonmetallic and Complex Substrates   60
   Appendix   61
   References   61

7 Influence of Ligands   65
   7.1 Phthalocyanine Molecules with Ligands   65
   References   67

8 Applications   69
   8.1 Spin Filtering Concept   69
   8.2 Molecular Multilayers/Molecular Films   70
   References   73

Acronyms

2PPE	Two-Photon Photoemission
ACFD	Adiabatic Connection Fluctuation Dissipation
ADF	Amsterdam Density Functional (code)
AF	Anti-Ferromagnetic
AS	Antibonding State
B3LYP	Becke, 3-parameter, Lee–Yang–Par (hybrid functional)
BS	Bonding State
CCI	Constant Current Imaging
CHI	Constant Height Imaging
CTQMC	Continuous Time Quantum Monte Carlo
D2	Grimme van der Waals approximation type 2
DC	Double Counting
DFT	Density Functional Theory
DOS	Density of States
DV	Discrete Variational
ED	Exact Diagonalization
ESCA	Electron Spectroscopy for Chemical Analysis
FHI-aims	Fritz-Haber-Institute-Ab Initio Molecular Simulations (code)
FLAPW	Full=potential Linearized-Augmented-Plane-Waves
FLL	Fully-Localized Limit
FM	Ferromagnetic
GGA	Generalized Gradient Approximation
Gs	Ground State
GTO	Gaussian-type Orbital
GW	Green function G—Coulomb interaction W
HF	Hartree–Fock
HOMO	Highest occupied molecular orbital
HS	High-Spin
HSE	Heyd–Scuseria–Ernzerhof (hybrid functional)
IETS	Inelastic Electron Tunneling Spectroscopy

KS	Kohn-Sham
LDA	Local Density Approximation
LDOS	Local Density of States
LEED	Low Energy Electron Diffraction
LS	Low-Spin
LUMO	Lowest unoccupied molecular orbital
MAE	Magnetocrystalline anisotropy Energy
ML	Monolayer
NEXAFS	Near Edge X-ray Absorption Fine Structure
NM	Nonmagnetic
PBE	Perdew–Burke–Ernzerhof (GGA version)
PBE0	Perdew–Burke–Ernzerhof (hybrid functional)
PBEh	Revised PBE
PES	Photo Electron Spectroscopy
PW	Plane Wave
PW91	Perdew-Wang 91 (GGA)
RPES	Resonant Photoelectron Spectroscopy
SIE	Self-Interaction Error
SOC	Spin-orbit Coupling
SOMO	Single Occupied Molecular Orbital
SP-STM	Spin-Polarized Scanning Tunneling microscopy
SQUID	Superconducting Quantum Interference Device
STM	Scanning Tunneling microscopy
StoBe	Stockholm-Berlin version of deMon (code)
TDDFT	Time-Dependent Density Functional Theory
TM	Transition Metal
TMPc	Transition Metal Phthalocyanine
U	Hubbard U parameter
UHFS	Unrestricted Hartree Fock Slater
UHV	Ultra-High Vacuum
UPS	Ultraviolet Photoelectron Spectroscopy
VASP	Vienna Ab-initio Simulation Package (code)
VB	Valence Band
vdW	van-der Waals
XAS	X-ray Absorption Spectroscopy
XMCD	X-ray Magnetic Circular Dichroism
XPS	X-ray Photoelectron Spectroscopy
XSW	X-ray Standing Waves
ZORA	Zeroth Order Regular Approximation (to Dirac equation)

Chapter 1
Introduction

Abstract Phthalocyanine molecules with a 3d transition metal in the centre, like MnPc, FePc, CoPc, NiPc and CuPc, have attracted a huge interest in the last decades due to the large number of possible applications. Experimental and theoretical gas phase studies are an important reference to understand the properties of the molecules, as well as how they can be modified and manipulated upon deposition on substrates or in supramolecular conformations. However, in several 3d metal phthalocyanines, the electronic structure of the single molecule is still under debate even after several spectroscopic studies and computational works have been performed. This is mostly due to the highly correlated 3d electrons of the metal atoms, which pose a challenge for the theory. In addition, the experiments to determine the electronic structure are often carried out in different conditions (on thick films or in gas phase for example), and this can lead to different results. The following chapter provides an overview of the theoretical and experimental results and debates related to the electronic structure of gas phase MnPc, FePc, CoPc, NiPc and CuPc.

Phthalocyanine molecules and complexes are currently investigated with a wide range of experimental and theoretical tools. Both optical and electronic properties are of interest, either from a fundamental point of view or having applications in mind. What makes these molecules interesting is the fact that their magnetic properties can be varied by adding ligand, changing the metal center, the substrate or possible end groups. These tunable magnetic properties make them candidates for various applications [1–8]. The phthalocyanines (Pc) are e.g. studied, in parallel to the related porphyrins, as organic dyes in dye-sensitized solar cells [9–12]. Their potential use in electrochromic displays and sensors [13] and gas sensors [14] have also been investigated, as well as their magnetic properties and the potential for spintronic applications [15]. Special attention has been paid to hybrid systems of molecules and substrates as possible building blocks for future electronic and magnetic devices [16–18]. In view of such devices, phthalocyanine molecules are advantageous because they adsorb—under favourable conditions—flat on metallic or semiconducting substrates [16, 19–21]. CuPc and MnPc molecules are discussed for optical applications e.g. solar cells or transistors [22, 23] whereas Fe-, Co- and CuPc molecules are interesting for spintronic applications [2, 4, 16]. SnPc and VOPc molecules are exceptional cases because they are not flat as the other TM phthalocyanines but also

H. C. Herper et al., *Molecular Nanomagnets*, Nanoscience
and Nanotechnology, https://doi.org/10.1007/978-981-15-3719-6_1

Fig. 1.1 Sketch of a
transition metal (TM)
phthalocyanine molecule.
Five membered pyrrole rings
(C and N atoms) surround
the central TM ion which is
bonded to four N atoms
(N_1). The *aza*-N (N_2) do not
share a direct bond with the
TM center

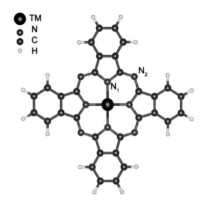

discussed in view of electronic devices [21, 24]. Furthermore the phthalocyanines
and porphyrins are of particular experimental interest to investigate due to their high
thermal stability, and the fact that they can be investigated both in gas phase as well as
films and monolayer to sub-monolayer coverages on a wide range of substrates [25–
27]. Both metallic, semiconducting and insulating substrates have been considered
in this context [28, 37].

In the group of phthalocyanines, the transition metal (TM) based molecules, where
the transition metal atom is Mn, Fe, Co, Ni or Cu, are of particular interest, since
the coupling between magnetic properties and electronic configuration is typically
strong [29, 30, 37], and in general not well understood. In Fig. 1.1 as an example
a TM based phthalocyanine (TMPc) molecule is shown. The TM atom sits in the
center of the molecule, and is bonded to four N atoms. These atoms are in turn part
of five membered pyrrole rings (involving C and N) that couple to six membered
benzene rings (C and H atoms). Transition metal phthalocyanine (TMPc) molecules
with TM being Mn, Fe, Co, Ni, or Cu are in general flat in gas-phase and have in
most cases D_{4h} or D_{2h} symmetry. The transition metal in this molecule has a formal
charge of $+2$. However hybridization between ligand orbitals and possible substrate
atoms, distorts this simple picture to various degrees. In these molecules the TM
atom normally has a finite magnetic moment, but the value of this moment depends
on the TM atom and how the molecule interacts with its surrounding, e.g. if it is
placed on a substrate. However, if deposited on a magnetic substrate not only the
magnetic moments can vary but in addition the magnetic coupling to the substrate
can be manipulated by ligands or substrate modification [31]. The magnetic moment
of the TM atom can also be made to couple ferromagnetically or antiferromagnetically to magnetic substrate atoms, depending on substrate. This possibility to tune
the magnetic properties of the molecule by 'external means' is attractive for applications. Among the experimental methods that have been used to study these molecules,
one should mention spectroscopic tools like ultra-violet photoelectron X-ray spectroscopy (UPS) [32], X-ray photoelectron spectroscopy (XPS) [33], X-ray absorption
spectroscopy (XAS) [34], that all aim to investigate the electronic structures of occupied states (UPS and XPS) and unoccupied states (XAS). The structural properties of

these molecules have primarily been investigated by scanning tip microscopy (STM) [35] and low energy electron diffraction (LEED) [36]. The magnetic properties have primarily been investigated using SQUID magnetometry [37] and X-ray magnetic circular dichroism (XMCD) [38]. As mentioned the coupling between electronic and magnetic properties is in general strong in these systems, as is the coupling to structural properties when they are adsorbed on a substrate. Since various investigations have focused on different substrates, film thicknesses or gas-phase information, it has been difficult to form a general consensus on the magnetic properties and the electronic configuration of these systems. From the theoretical side, first principles calculations of ground state electronic configurations as well as spectra have been performed using LSDA, LSDA+U as well as hybrid functionals [22, 39, 40]. In addition, these methods have been used to calculate the ground state spin-configuration, as well as the coupling to a substrate have been investigated. However, also from theory it is sometimes difficult to reach a firm consensus of the properties of TMPc's and to reconcile theoretical data with experimental observations. This review outlines the current knowledge state of affairs of magnetism, electronic structure and geometry of TMPc's in gas phase as well as in film form, both as concerns the experiments as well as from a theoretical point of view and tries to reveal magnetic trends depending on the choice of the TM atom and the substrate.

References

1. Sedona F, Marino MD, Forrer D, Vittadini A, Casarin M, Cossaro A, Floreano L, Verdini A, Sambi M (2012) Nat Mater 11:970
2. Heutz S, Mitra C, Wu W, Fisher A, Kerridge A, Stoneham M, Harker AH, Gardener J, Tseng HH, Jones T, Renner C, Aeppli G (2007) Adv Mater 19
3. Xhen S, Lu W, Yao Y, Chen H, Chen W (2014) J React Kinet Mech Catal 111:535
4. Huang J, Xu K, Lei S, Su H, Yang S, Li Q, Yang J (2012) J Chem Phys 136:064707
5. Yang R, Gredig T, Colesniuc C, Schuller IK, Park J, Trogler BC, Kummel AC (2007) Appl Phys Lett 90:263506
6. Bohrer FI, Colesniuc CN, Park J, Ruidiaz ME, Schuller IK, Kummel A, Trogler WC (2009) J Am Chem Soc 13:478
7. Axtabal A, Ribeiro M, Parui S, Urreta L, Sagasta E, Sun X, Llopis R, Casanova F, Hueso LE (2016) Nat Commun 7:13751
8. Niu Li, Wang Huan, Bai Lina, Rong Ximing, Liu Xiaojie, Li Hua, Yin Haitao (2017) Front Phys 12:127207
9. de la Torre G, Claessens CG, Torres T (2007) Chem Commun 20:2000
10. J. Mack, N. Kobayashi, Chem. Rev. (Washington, DC, U.S.) **111**, 281 (2011)
11. Claessens CG, Hahn U, Torres TT (2008) Chem Rec 8:75–97
12. Chem J (2014) Phys 118:17166
13. (2009) J Porphyr Phthalocyanines 13:606
14. Muzikante I, Parra V, Dobulans R, Fonavs E, Latvels J, Bouvet M (2007) Sensors 7:2984
15. Iacovita C, Rastei MV, Heinrich BW, Brumme T, Kortus J, Limot L, Bucher JP (2008) Phys Rev Lett 101:116602
16. Klar D, Klyatskaya S, Candini A, Krumme B, Kummer K, Ohresser P, Corradini V, de Renzi V, Biagi R, Joly L, Kappler JP, del Pennino U, Affronte M, Wende H, Ruben M (2013) Beilstein J Nanotechnol 4:320

17. Xiong ZH, Wu D, Vardeny ZV, Shi J (2004) Nature 427:821
18. Cai YL, Rehman RA, Ke W, Zhang HJ, He P, Bao S (2013) Chem Phys Lett 582:90
19. Sun X, Wang B, Yamauchi Y (2012) J Phys Chem C 116:18752
20. Guo Q, Huag M, Qin Z, Cao G (2012) Ultramicroscopy 118:17
21. Mattioli G, Filippone F, Bonapasta AA (2006) J Phys Chem Lett 1:2757
22. Brumboiu I, Totani R, de Simone M, Coreno M, Grazioli C, Lozzi L, Herper HC, Sanyal B, Eriksson O, Puglia C, Brena B (2014) J Phys Chem A 118:927
23. Dimitrakopoulos C, Mascaro DJ (2001) IBM J Res Dev 45:11
24. Lackinger M, Hietschold M (2002) Surf Sci Lett 520:L619
25. Ishikawa N (2010) vol 135, p 211. Springer
26. Delacote G, Fillard JP, Marco FJ (1964) Solid State Commun 12
27. Li W, Yu A, Higgins D, Llanos B, Chen Z (2010) J Am Chem Soc 132:17056
28. Ponce I, Francisco S, Silva J, Onate R (2011) J Phys Chem C 115:23512
29. Klar D, Brena B, Herper HC, Bhandary S, Weis C, Krumme B, Schmitz-Antoniak C, Sanyal B, Eriksson O, Wende H (2013) Phys Rev B 88:224424
30. Brena B, Herper HC (2015) J Appl Phys 117:17B318
31. Herper HC, Bhandary S, Eriksson O, Sanyal B, Brena B (2014) Phys Rev B 89:085411. https://doi.org/10.1103/PhysRevB.89.085411
32. Nardi MV, Detto F, Aversa L, Verucchi R, Salviati G, Iannotta S, Casarin M (2013) Phys Chem Chem Phys 15:12864
33. Chem J (2012) Phys 137:054306
34. Willey T, Bagge-Hansen M, Lee JR (2013) J Chem Phys 139:034701
35. Mugarza A, Robles R, Krull C, Korytár R, Lorente N, Gambardella P (2012) Phys Rev B 85:155437
36. (2014) Surf Sci 621. https://doi.org/10.1016/j.susc.2013.10.020
37. Yamada H, Shimada T, Koma A (1998) J Chem Phys 108:10256
38. Candini A, Bellini V, Klar D, Corradini V, Biagi R, Renzi VD, Kummer K, Brookes NB, del Pennino U, Wende H, Affronte M (2014) J Phys Chem C 118
39. Ali ME, Sanyal B, Oppeneer PM (2009) J Phys Chem C 113:14391
40. Ahlund J, Nilson K, Schiessling J, Kjeldgaard L, Berner S, Mårtensson N, Puglia C, Brena B, Nyberg M, Luo Y (2006) J Chem Phys 125:34709

Chapter 2
Experimental Techniques

Abstract Different spectroscopic methods can be used to characterize the electronic structure of a system of interest, i.e. molecular, solid or adsorbate samples. The different techniques give complementary information about the geometric and electronic structure of the system. By Photoelectron Spectroscopy (PES), Auger and resonant photoemission (RPES) the occupied electronic levels can be studied, whereas X-ray Absorption Spectroscopy (XAS) gives information about the unoccupied valence states of the system in presence, however, of a core hole. Magnetic information can be obtained from X-ray Magnetic Circular Dichroism (XMCD).

2.1 X-ray Absorption Spectroscopy (XAS)

The attenuation of electromagnetic radiation in matter is described by the attenuation coefficient $\tilde{\mu}(E)$ which is given by Beer's law:

$$I(x) = I_0 \, e^{-\tilde{\mu} x} . \tag{2.1}$$

Here, $I(x)$ is the intensity of X-rays after the primary intensity I_0 is attenuated by a sample with thickness x. The elastic Rayleigh and the inelastic Compton scattering can be neglected in the X-ray regime investigated here. Hence, the attenuation coefficient $\tilde{\mu}$ is approximately identical to the photoelectric absorption coefficient μ. This is the key property in the X-ray absorption spectroscopy. Characteristic features of the X-ray absorption coefficient are the absorption edges: A step-like increase is found in the X-ray absorption coefficient when the photon energy is large enough to excite an atomic core electron into the continuum. These edges are labeled according to the Sommerfeld notation: edges that stem from excitation of the core electrons $1s_{1/2}, 2s_{1/2}, 2p_{1/2}$ and $2p_{3/2}$ are labelled as K, L_1, L_2 and L_3 edges, respectively. The binding energies of the electrons are characteristic for the respective element. Hence, the analysis of specific absorptions edges makes the X-ray absorption spectroscopy an element specific tool to study e.g. the electronic structure or the magnetic properties of the elements in the sample. The absorption coefficient defined above thus

H. C. Herper et al., *Molecular Nanomagnets*, Nanoscience
and Nanotechnology, https://doi.org/10.1007/978-981-15-3719-6_2

is a photon energy dependent property $\mu(h\nu)$. This energy dependence in the vicinity of the absorption edge can described by Fermi's golden rule in the one-electron approximation:

$$\mu(h\nu) \propto \sum_{i \, occ} \sum_{f \, unocc} |\langle \psi_f | \mathbf{p} \cdot \mathbf{A}(\mathbf{r}) | \psi_i \rangle|^2 \cdot \rho(E_f) \cdot \delta(E_f - E_i - h\nu), \quad (2.2)$$

where $|\psi_f\rangle$ and $|\psi_i\rangle$ are the (unoccupied) final and (occupied) initial states with energies E_f and E_i, respectively. If the fine structure of a single absorption edge is analyzed, only the sum of the unoccupied final states $|\psi_f\rangle$ enters into Eq. 2.2 for a specific core shell. The absorption coefficient is linked to the angular momentum projected density of unoccupied states which is seen in Eq. 2.2 in the density of final states $\rho(E_f)$. The interaction operator is given by $\mathbf{p} \cdot \mathbf{A}(\mathbf{r})$ describing the interaction of the electron (with momentum operator $\mathbf{p}$) with the electromagnetic field (given by the vector potential $\mathbf{A}(\mathbf{r})$). The vector potential $\mathbf{A}(\mathbf{r})$ here is a classical wave with the polarization $\boldsymbol{\varepsilon}$:

$$\mathbf{A}(\mathbf{r}) \cong \boldsymbol{\varepsilon} \cdot A_0 \cdot e^{i\mathbf{k}\cdot\mathbf{r}}. \quad (2.3)$$

As we will focus on the soft X-ray regime ($h\nu \approx 100\,\text{eV}–5\,\text{keV}$) we will only consider electric dipole transitions (E1) by approximating in Eq. 2.3:

$$e^{i\mathbf{k}\cdot\mathbf{r}} \cong 1. \quad (2.4)$$

Examples will be given in the following for XAS at the K-edges ($1s \rightarrow 2p$ transitions) of light elements as e.g. N and C using linearly polarized X-rays together with the study of $3d$ transition elements at the $L_{2,3}$-edges (dominated by $2p \rightarrow 3d$ transitions) utilizing circularly polarized light to analyze the magnetic properties. The element specific spin and orbital moments μ_S and μ_L of the $3d$ transition metals like Fe and Co can be determined by the sum rules as discussed by Thole et al. [1] and Carra et al. [2]:

$$\frac{\mu_L}{\mu_B} = -\frac{2N_h}{N} \int (\Delta\mu_{L_3} + \Delta\mu_{L_2}) dE \quad (2.5)$$

$$\frac{\mu_S}{\mu_B} = -\frac{3N_h}{N} \int (\Delta\mu_{L_3} - 2\Delta\mu_{L_2}) dE + 7\langle T_{zz}\rangle \quad (2.6)$$

The X-ray magnetic circular dichroism (XMCD) signal is described here e.g. at the L_3 edge by the difference of the X-ray absorption coefficients for right and left circularly polarized X-rays $\Delta\mu_{L_3} = \mu_{L_3}^+ - \mu_{L_3}^-$. In Eqs. 2.5 and 2.6 the data are normalized by the integrated spectrum for unpolarized radiation:

$$N = \int_{L_3+L_2} (\mu^+ + \mu^- + \mu^0) dE \quad (2.7)$$

Furthermore, the number of unoccupied d-states N_h enters in Eqs. 2.5 and 2.6 because of the following reason: Exciting the $2p_{1/2}$ and $2p_{3/2}$ electrons by circularly polarized X-rays leads to a spin-polarization of the photoelectron. If the $3d$ system is magnetic i.e. if the unoccupied $3d$ states are spin split the final states act as a sensor for the spin-polarized photoelectrons. By integrating the XMCD, the imbalance of the unoccupied states can be detected. If the total number of holes in the d states N_h is known, this signal can be used to draw conclusions on the occupied $3d$ states, and thereby the magnetic moments can be determined by integrating the XMCD signal. However, care has to be taken of the magnetic dipole term $\langle T_Z \rangle$ which enters into the spin sum rule in Eq. 2.6. This term is the expectation value of the z-component of the magnetic dipole operator

$$\mathbf{T} = \mathbf{s} - \frac{3\mathbf{r}(\mathbf{r} \cdot \mathbf{s})}{r^2} \tag{2.8}$$

which describes the asphericity of the spin magnetization. Actually $\langle T_z \rangle$ can be quite large for magnetic molecules [3]. In the next section the XAS and XMCD spectra are discussed for the example case FePc deposited on a Co film with and without an oxygen adlayer on the metal substrate.

2.1.1 Coupling of FePc to Co Film

As a single molecule Fe-phthalocyanines (FePc) behave paramagnetically. However, for possible molecular spintronics applications the ordering of the individual magnetic moments is necessary at room temperature. Since this ordering cannot be achieved at 300 K for paramagnetic systems with technical feasible magnetic fields in the regime of 1T, a different approach is necessary. It has been demonstrated for Mn-porphyrin and Fe-porphyrin molecules that the molecular spins of the $3d$ centers can be ordered by direct and indirect exchange coupling to ferromagnetic films [4, 5]. Furthermore, the magnetic coupling of the $3d$ atoms in the molecules to the ferromagnetic films can be tailored by means of an intermediate layer of atomic oxygen [6]. The coupling changes from ferromagnetic for the case without the intermediate oxygen layer to antiferromagnetic when utilizing the exchange coupling via the oxygen atoms. The X-ray absorption spectroscopy and especially the X-ray magnetic circular dichroism are ideally suited techniques to elucidate the interaction in the molecular hybrid systems. This will be demonstrated in the following for the case of Fe-phthalocyanine molecules adsorbed on epitaxial Co films on a Cu(001) single crystal [7]. The schematic presentation of the molecular hybrid system and the measuring geometry is presented in Fig. 2.1. A 5 ML Co film is prepared in situ directly at the synchrotron radiation facility (BESSY II) on the Cu(001) crystal by means of electron-beam evaporation under UHV conditions. The submonolayer of the Fe(II)Pc molecules were sublimated from a Knudsen cell onto the substrate held at room temperature. As presented in Fig. 2.2a a ferromagnetic coupling is found when

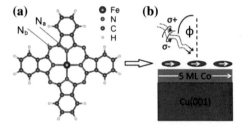

Fig. 2.1 a Sketch of the FePc molecule which is composed of an Fe (purple) ion in the center, 8 N (gray), 32 C (blue), and 16 H (yellow) atoms. **b** Schematic illustration of the XMCD measurement for the sample FePc/Co/Cu(001) under the incidence angle ϕ [7]

Fig. 2.2 Fe $L_{2,3}$ XAS and XMCD spectra of 0.8 ML FePc/5 ML Co (**a**), 0.8 ML FePc/O/5 ML Co (**b**) measured at T = 32 K and with a grazing angle (ϕ) of 70° [7]

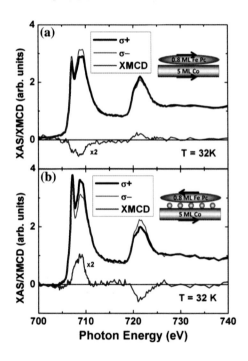

the FePc molecules are adsorbed directly on the Co film. To achieve an antiparallel orientation an intermediate layer of atomic oxygen can be used. This is prepared by dosing oxygen on the clean Cu(001) crystal. Thereby a reconstruction with an atomic oxygen $(\sqrt{2} \times 2\sqrt{2})R45°$ superstructure is created, [7]. When growing the Co film on this superstructure the oxygen atoms act as surfactants yielding a $c(2 \times 2)$ superstructure of 0.5 ML atomic oxygen on top of the Co film [7]. This system is then used as the template for the preparation of the molecules. That indeed an anti-ferromagnetic coupling is achieved is seen in Fig. 2.2b where the sign of the XMCD is changed as compared to the case without the intermediate layer of oxygen.

To get further insight into the interaction of the molecules with the substrate the nitrogen K-edge was studied with linearly polarized X-rays. By utilizing the so-called 'search-light' effect, the study of the angular dependence of the NEXAFS data with linearly polarized X-rays yields information on the orientation of the molecules on the surface. At the K-edge transitions from the initial $1s$ to final p states are probed by electric dipolar transitions (E1). For molecular systems this has the consequence that the antibonding σ^* molecular orbitals are analyzed if the electric field vector is oriented in the molecular plane. However, excitations to the antibonding π^* orbitals are studied if the electric field vector is aligned perpendicular to the molecular plane. In addition, the modification/broadening of fine structures especially at the π^* resonances provides information on the interaction of the molecules with the substrate. The experimental angular dependent NEXAFS spectra for the FePc/Co system are shown in Fig. 2.3 for grazing ($\phi = 70°$) and normal X-ray incidence ($\phi = 0°$) together with calculated spectra using the DFT code StoBe [8]. At first the experimental spectra shall be discussed: A pronounced angular dependence of the data is determined which indicates that the molecules adsorb flat on the surface since the σ^* resonance at 405–425 eV is detected at normal X-ray incidence (Fig. 2.3b) with the E-vector parallel to the surface, whereas the π^* resonances at 398–405 eV are seen at grazing X-ray incidence (Fig. 2.3a) with the E-vector essentially out of plane. These experimental results are compared to the DFT calculations for the molecules with D_{4h} symmetry (gas phase molecule) and the relaxed structure for the adsorbed molecule. The direct comparison between these two calculations and the experimental data reveals that the calculations using the relaxed structure describe

Fig. 2.3 Nitrogen K-edge experimental and theoretical spectra. **a** quasi perpendicular (70°). **b** parallel (0°) polarization. The experimental spectrum for the case without an intermediate layer of atomic oxygen is shown as black thick line, the black thin line is the total calculated spectrum for the relaxed FePc and the red line is the total spectrum for the symmetric FePc. The bar graphs show the calculated oscillator strengths (red bars represent the symmetric FePc and black bars the relaxed FePc) [7]

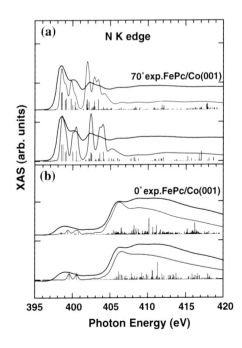

the experimental results especially in the regime of the π^* resonances better. As can be seen e.g. for the data at normal X-ray incidence (Fig. 2.3b) a partial out-of-plane distortion of the FePc molecule is found for the relaxed structure. This has the consequence that the π^* orbitals are not perfectly aligned along the out-of-plane direction which results in contributions that are also visible when the E-vector is parallel to the surface (normal X-ray incidence). Therefore the remainder of the π^* resonances at normal X-ray incidence is an indication of a relatively strong interaction of the FePc molecules with the Co surface (Fig. 2.4).

The DFT calculations even allow for a disentanglement of the contributions to the π^* resonances of the inequivalent nitrogen atoms N_a and N_b as shown in Fig. 2.1. As the nitrogen atom N_a is bound to the iron atom and to two carbon atoms the contributions of this atom is located at higher photon energies as compared to the nitrogen atom N_b which is bound to two carbon atoms only. The fine structures in the NEXAFS spectra can be related to the isodensity surfaces for the final-state orbitals as calculated by StoBe [8]. Thereby the fine structures can be directly connected to the electronic structure of the adsorbed molecules on the Co surfaces. As shown by Klar et al. the intermediate layer of atomic oxygen effectively decouples the nitrogen π system from the surface resulting in NEXAFS fine structures at the π^* resonances which resemble more gas-phase-like structures [7]. This indicates that the FePc molecules on the oxygen covered Co surface maintains a flat structure and the interaction is mainly mediated by the Fe atom in the molecule. This can be nicely seen in Fig. 6.3 where the magnetic density for the two systems is shown.

Fig. 2.4 Nitrogen K-edge, detail. **a** simulation of perpendicular polarization for N_a; **b** simulation of perpendicular polarization for N_b. The experimental spectrum in the quasi perpendicular set up at 70° is shown (grey thick line). The red dashed line is the total spectrum for the relaxed FePc. The bar graphs show the calculated oscillator strengths for N_a (**a**) and N_b (**b**) [7]

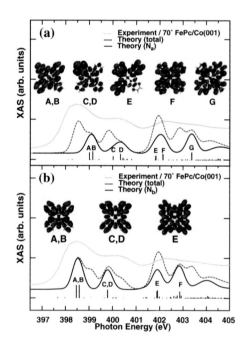

2.2 Photoelectron Spectroscopy (PES)

In photoelectron spectroscopy (PES) a sample is exposed to photons of sufficient high energy to ionize the atoms (or the molecules) of the system emitting electrons. This process, illustrated in Fig. 2.5, and called photoemission, is based on the photo-electric effect and the emitted electrons are called photoelectrons. In photoemission spectroscopy, the kinetic energy of the photoelectrons is measured. The conservation of energy implies that the photon energy ($h\nu$), the initial binding energy (E_B) of the photoelectron (prior to photoemission) and its final kinetic energy (E_K) are related by the equation:

$$E_B = h\nu - E - \varPhi = E_{final} - E_{initial} \tag{2.9}$$

where $\varPhi$ represents the work function of the system. The binding energy in this way is related to the vacuum level (E_V). Usually in surface science, the binding energies are referred to the Fermi level (E_F) and then the term $\varPhi$ in the equation can be neglected. The relation in Eq. 2.9 is the basis of the photoemission spectroscopy because, knowing $h\nu$ and E_K, it permits the determination of the binding energy of the emitted electrons, which appear as intensity peaks at certain binding energies in the spectrum. In more detail, a photoelectron spectrum usually consists of the main line and other extra structures, which usually appear at higher binding energy, i.e. the so called satellites. They are the results of the perturbation of the electronic structure during the photoelectron process. They can be considered as excitations of valence electrons to bound or continuum states, called shake-up or shake-off respectively. Since these excitations decrease the kinetic energy of the outgoing electron the satellites appear at higher binding energy than the main line. The photoemission process is very fast (10^{-17} s) and the core hole final state is usually a non-equilibrium state, which often leads to vibrational excitations. The vibrational energy and the width of the vibrational states can result in separated peaks or in a broadening of the main line.

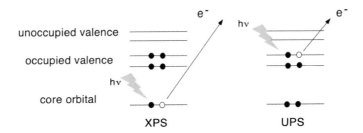

Fig. 2.5 Sketch of core and valence photoemission processes. X- ray photons are needed for studying the core levels of a system and the technique is then called X-ray Photoelectron Spectroscopy (XPS). Ultra-violet Photoelectron Spectroscopy (UPS) is the name of the same method when ultraviolet light is used for a detailed characterization of the valence levels

Since each element in the periodic table is characterized by a specific set of core and valence level binding energies, photoemission spectroscopy is a powerful tool to analyze the chemical composition of a sample and hence is also called Electron Spectroscopy for Chemical Analysis (ESCA). This technique is also used for determining the amount of chemical species, since the intensity of a photoemission peak is related to the number of emitted photoelectrons with certain kinetic/binding energy.

In order to probe the core electrons, characterized by a high binding energy, it is necessary to use X-ray photons ($h\nu \geq 1000$ eV) and the technique is called X-ray Photoelectron Spectroscopy (XPS).

The core levels of an atom or of a molecule do not participate in the bonding with other atoms. However, if the environment of the studied atom changes, e.g., due to the adsorption on a surface, this causes a shift of the specific core levels. The existence of the chemical shift represents the most important aspect of the core electron spectroscopy and it can be used as fingerprint of chemical species of different molecules/complex or even chemically non-equivalent atoms in the same molecule [9]. For adsorbates, the chemical shift permits to distinguish different chemical species on the same surface, it gives information on the electronic structure and the chemical stability of the systems. Moreover, a comparison of data of the same molecules in gas phase and as adsorbate can provide insights in the modification induced by adsorption on the molecular electronic structure. Since as shown in Eq. 2.9 the binding energy of a photoelectron can be expressed as the energy difference between the initial (N electrons) state and the final (N-1 electrons) state:

$$E_B = E_{\text{final}}(N-1) - E_{\text{initial}}(N) \quad . \tag{2.10}$$

the chemical shift between two systems, a and b, can be written as the difference in total energies related to the initial and to the final states :

$$\Delta E = E_{B,a} - E_{B,b} \tag{2.11}$$
$$= [E_{\text{final,a}}(N-1) - E_{\text{initial,a}}(N)] - [E_{\text{final,b}}(N-1) - E_{\text{initial,b}}(N)]$$
$$= \Delta E_{a,b}(\text{final}) - \Delta E_{a,b}(\text{initial}) \quad .$$

Each of these two terms can be dominant for different systems. If we consider for example, the case of chemically inequivalent atoms in the same system, the binding energy shift can be attributed to the difference in the total energy of the two final states [10] since they have the same initial state energy. One way to estimate the contributions to the chemical shift is to us a Born–Haber model [11] relating the ionization process to measurable thermodynamic quantities.

The valence orbitals are the outermost levels of the atoms (or molecules) and therefore they have a rather low binding energy. Ultra-violet photons (of about 10–40 eV) can be used to study the valence electrons and this method is called Ultra-Violet Photoelectron Spectroscopy (UPS). The valence levels can be energetically and spatially modified by the bonding and they can form delocalized bands as in metals.

The photoionization cross section for each electron level, i.e. the probability for the ionization of each level, depends on the photon energy and varies strongly for different core levels. This means that in order to optimize the intensity of a core or valence electron spectrum, it is necessary to have a tunable photon source, i.e. the synchrotron radiation. By choosing different photon energies it is also possible to enhance the photoemission from an adsorbate orbital with respect to the emission from substrate levels which occurs within the same binding energy region. Furthermore, the tunability of synchrotron radiation permits to change the surface sensitivity of the photoemitted electrons. This makes it possible to carry out experiments where bulk and surface components of a spectral feature (in a spectrum) can be distinguished. This can be used to estimate the coverage and the geometrical arrangement of an adsorbate (see Pt $4f$ spectra for the different oxygen adsorption phases in Ref. [12]). Moreover in different studies, the variation of the cross section has been used for identifying the character of the outermost valence spectrum peaks, i.e. the character of the HOMO structure [13]. Such studies have also been dedicated to investigate the valence structure of Phthalocyanines [14].

2.3   Adsorbate Core Level Shift: Final State Effects

Upon adsorption of gas atoms or molecules on a surface, new relaxation (screening) channels of the core hole final state are available with respect to the case of free atoms (molecules). The surface can induce an extra screening channel, which lowers the final state energy by several eV. Many studies in the past have been dedicated to the characterization of the different kind of adsorption of gas molecules on solid surfaces. Depending on the adsorption strength, it is common to distinguish the adsorption in chemisorption and physisorption even if they are not so rigid and defined concepts. Depending on the strength of the adsorption we usually distinguish two main screening processes: metallic screening (for chemisorbed systems) and image potential screening (for physisorption systems).

In a chemisorption process, a chemical bond is formed between the adsorbate and the surface, producing often drastic changes in the electronic structure of the adsorbed molecules. In many cases, this leads to the formation of new molecular phases or dissociation upon the adsorption on a substrate. Due to the adsorbate-surface orbital overlap, a charge transfer from the substrate to the adsorbate, leading to a neutral final state, can screen the core hole state produced by photoionization.

For physisorbed systems the interaction between the adsorbate and the surface is very weak and it is mediated mainly via van der Waals forces. The overlap between the valence orbitals of the adsorbate and the surface is very small. The physisorbate can be considered as electronic isolated from the substrate and it preserves many gas-like properties. There is no charge transfer screening of the core hole produced in the ionization process. However, the core hole energy is lowered due to the creation of an image charge in the substrate and to the polarization of the neighbouring molecules.

An important and unique aspect of the core and valence level spectroscopies such as XAS, Auger and Resonant photoemission spectroscopies, is the possibility to locally probe the electronic structure of the adsorbate-substrate system. Due to the localization of the core level the different spectroscopies give information regarding the local density of state (electronic structure) at the core hole site. This implies that even if the core electrons do not participate in the chemical bonds with the substrate nevertheless the core level spectroscopy methods provide information about the chemistry which involves the valence levels. However, in order to correctly interpret the observed shifts, different theoretical treatments can be used described in the following chapters of this review.

2.4 Scanning Tunneling Microscopy (STM)

Developed in 1982 by Binnig and Roher [15, 16], the Scanning Tunneling Microscopy (STM) was soon recognized as one of the most powerful surface characterization method and awarded by the Nobel Prize in 1986. In fact it allows imaging and characterization of surfaces and surface structures at atomic resolution, which are strongly related to the surface selectivity and reactivity. STM is also a very flexible technique which can be used in a wide range of pressure, i.e. from UHV to high pressure. Moreover, giving image of surface reconstruction, adsorption structure/geometries working under variable temperature conditions, STM allows to study the dynamics during surface reactions [17–19].

A scanning tunneling microscope uses a sharp (mono-atomic sharp [20]) metallic tip which approaches a surface at a distance of a few Å resulting in a such thin barrier that the tails of the electron wave functions of the tip and the surface can overlap, see Fig. 2.6a. An applied bias voltage, V_{bias} between the tip and the surface causes a misalignment of the Fermi levels, driving in this way electrons to tunnel between the surface and the tip though the barrier giving rise to a tunnelling current, I_t. Depending on the sign of the V_{bias}, the tunnelling current can pass from the tip to the surface or vice versa, providing information about the occupied or unoccupied density of states (DOS) close to the Fermi level. Since the technique is based on electron tunneling, both the surface and the tip should be conducting, i.e. metals or semiconducting material. The tunnelling current I_t can be expressed as:

$$I_t \propto e^{-2\frac{\sqrt{2m\phi}}{\hbar}d} \tag{2.12}$$

where d represents the distance between the tip and the sample surface (or barrier width), ϕ is the local barrier height and m is the mass of a free electron [21]. The equation shows that the tunnelling current is exponentially dependent on the variations of the barrier width d, which experimentally is tuned by the distance between the tip and the surface, i.e. z in the coordinate systems usually used. For each Å variation in z, the tunnelling current will vary of one order of magnitude. This extreme sensibility

Fig. 2.6 a A schematic illustration of the tunneling event between tip and sample. Filled areas show occupied DOS and barrier heights ϕ_s and ϕ_t are indicated, positive bias of eU_{bias} is applied to sample and electrons tunnel from tip to sample probing the empty DOS of sample; **b** a schematic figure of the STM, **c** principle of constant current imaging mode (CCI), **d** principle of constant height imaging mode (CHI)

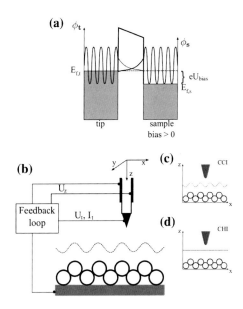

of the tunnelling current permits the measurement of surface corrugation at atomic scale. Of course such a performance requires high accuracy in z-movements (better than 0.1 Å), providing a fine positioning of the tip with respect to the sample surface. This is realized by using piezoelectric elements, which can change their dimensions in the order of an Å when a bias voltage is applied. Then, fundamental experimental components of a ST Microscope are the tip, the piezo-motors used for scanning, a proper vibration isolation and feedback loops of the instrument.

A schematic picture of the STM operating principle is shown in Fig. 2.6b. A STM image is obtained by scanning the tip parallel to the surface with piezoelements, which control also the x and y directions at sub-Å scale. There are two STM acquisition modes: the Constant Current Imaging (CCI) and Constant Height Imaging (CHI) [21, 22]. In CCI mode (Fig. 2.6c) the tunneling current between the tip and the sample surface is kept constant via a continuous adjustment of the tip height during the scanning of the surface. The data then are given as function of $z(x, y)$, monitoring the height changes of the tip for keeping the current constant. For high electron tunneling probability the tip retracts from the surface, whereas for low electron tunneling probability the tip moves closer to the surface.

If the Constant Height Imaging mode is used, the tip position is kept constant and the image is given from the variations of the tunneling current I_t while scanning the surface, see Fig. 2.6d.

Even if STM is considered giving the image of the surface, we should always consider that the contrast observed is actually the representation of the local density of states (LDOS) on the sample at the Fermi level convoluted with the probing state at the end of the tip. For a detailed description of the tunneling process, an accurate knowledge of the electronic states of the tip and of the substrate is needed.

However, the poor characterization of the tip, its unknown real shape and the unknown interaction between the surface and the tip put serious limitation to such an accurate description. For strong tip-substrate interactions, it would be necessary to consider the tip, the substrate and the vacuum barrier as one system instead of considering them as weakly interacting separated systems [20]. The interpretation of the STM data should always take into consideration the different tip-induced effects. Just to mention an example, a large electronic radius of the tip can drastically influence the image. Interacting with different adjacent surface atoms, it can cause an electronic interference between different tunneling channels, resulting then in an image that is far different from the real surface topography [23].

References

1. Thole BT, Carra P, Sette F, van der Laan G (1992) Phys Rev Lett 68:1943
2. Carra P, Thole BT, Altarelli M, Wang X (1993) Phys Rev Lett 70:694
3. Herper HC, Bernien M, Bhandary S, Hermanns CF, Krüger A, Miguel J, Weis C, Schmitz-Antoniak C, Krumme B, Bovenschen D, Tieg C, Sanyal B, Weschke E, Czekelius C, Kuch W, Wende H, Eriksson O (2013) Phys Rev B 87:174425 https://doi.org/10.1103/PhysRevB.87. 174425
4. Scheybal A, Ramsvik T, Bertschinger R, Putero M, Nolting F, Jung TA (2005) Chem Phys Lett 411:214
5. Wende H, Bernien M, Luo J, Sorg C, Ponpandian N, Kurde J, Miguel J, Piantek M, Xu X, Eckhold P, Kuch W, Baberschke K, Panchmatia PM, Sanyal B, Oppeneer PM, Eriksson O (2007) Nat Mater 6:516
6. Bernien M, Miguel J, Weis C, Ali ME, Kurde J, Krumme B, Panchmatia PM, Sanyal B, Piantek M, Srivastava P, Baberschke K, Oppeneer PM, Eriksson O, Kuch W, Wende H (2009) Phys Rev Lett 102:047202
7. Klar D, Brena B, Herper HC, Bhandary S, Weis C, Krumme B, Schmitz-Antoniak C, Sanyal B, Eriksson O, Wende H (2013) Phys Rev B 88:224424
8. Hermann K, Pettersson LGM, Casida ME, Daul C, Goursot A, Koester A, Proynov E, St-Amant A, Salahub DR, Contributing authors: Carravetta V, Duarte H, Friedrich C, Godbout N, Guan J, Jamorski C, Leboeuf M, Leetmaa M, Nyberg M, Patchkovskii S, Pedocchi L, Sim F, Triguero L, Vela A (2007) StoBe-deMon version 3.0
9. Brena B, Luo Y, Nyberg M, Carniato S, Nilson K, Åhlnd J, Mårtensson N, Siegbhan H, Puglia C (2014) Phys Rev B 70:195214; Mårtensson N, Sokolowski E, Svensson S (2014) J Elect Spectrosc Relat Phenom 193:27
10. Nisslon A, Tillborg H, Mårtensson N (1991) Phys Rev Lett 67:1015
11. Johansson B, Mårtensson N (1980) Phys Rev B 21:4427
12. Puglia C, Nilsson A, Hernnä s B, Karis O, Bennich P, Mårtensson N (1995) Surf Sci 342:119
13. Bennich P, Puglia C, Brühwiler PA, Nilsson A, Maxwell AJ, Sandell A, Mårtensson N, Rudolf P (1999) Phys Rev B 59:8292
14. Brena B, Puglia C, de Simone M, Coreno M, Tarafder K, Feyer V, Banerjee R, Göthelid E, Sanyal B, Oppeneer PM, Eriksson O (2011) J Chem Phys 134:074312
15. Binnig G, Rohrer H, Gerber C, Weibel E (1982) Phys Rev Lett 49:57
16. Binnig G, Rohrer H (1982) Helv Phys Acta 55:726
17. Land T, Michely T, Behm R, Hemminger J, Comsa G (1992) Surf Sci 264:261
18. Besenbacher F, Laegsgaard E, Nielsen LP, Ruan L, Stensgaard I (1994) J Vac Sci Technol B 12:1758
19. Gimzewski JK, Berndt R, Schlittler R (1991) J Vac Sci Technol B 9:897

20. Besenbacher F, Laegsgaard E, Nielsen LP, Ruan L, Stensgaard I (1996) Rep Prog Phys 59:1737
21. Wiesendanger R (1994) Scanning probe microscopy and spectroscopy, methods and applications. Cambridge University Press, Cambridge
22. Bai C (1995) Springer, Shangai
23. McIntyre BJ, Sautet P, Dunphy JC, Salmeron M, Samorjai GA (1994) J Vac Sci Technol B 12:1751

Chapter 3
Theoretical Methods

Abstract In this chapter, the theoretical methods required for the description of structure, electronic structure and magnetism of magnetic molecules in the gas phase and in the adsorbed configurations will be discussed. The main workhorse of the theoretical methods is the density functional theory that provides a materials-specific description of electronic structure, which is quite sufficient for many of the materials. However, in the present context of magnetic molecules, one needs to go beyond standard approximations in density functional theory. In this regard, some of the crucial characteristics in the electronic structure and magnetism will be discussed such as electron correlation, van der Waals interaction, band gaps, magnetic anisotropy and spin-dipole moments.

3.1 Density Functional Theory

It is widely recognized nowadays that Density Functional Theory (DFT) is a very powerful method to calculate properties of materials from zero to three dimension with a good quantitative accuracy and predictive power. This theory is being regularly applied in the field of physics, chemistry and biology for tackling complicated problems. The heart of this theory is in the total energy expressed as a functional of electron density giving rise to an effective Hamiltonian [1] that has to be treated for the self-consistent solution of a single particle equation [2]. A number of review articles [3–5] have been published through years and the interested readers are referred to those.

3.2 Hybrid Functionals

Despite an immense success of DFT within the local and semilocal approximations of exchange-correlation functionals, it suffers from several drawbacks. One of them is the severe underestimation of HOMO-LUMO gaps in molecules. A commonly used approach taken by quantum chemists to deal with molecules is the hybrid

H. C. Herper et al., *Molecular Nanomagnets*, Nanoscience
and Nanotechnology, https://doi.org/10.1007/978-981-15-3719-6_3

functional theory where one combines Hartree–Fock (HF) and Density Functional Theory (DFT) to treat the exchange correlation part. This hybrid method contains a fraction of Hartree–Fock exchange defined in terms of single particle orbitals and the rest by DFT as a functional of electron density. The non-local Fock exchange in Hartree–Fock theory has the following form

$$E_X = -\frac{1}{2} \sum_{i,j} \int \int \phi_i^*(\mathbf{r}) \phi_j^*(\mathbf{r}') \frac{1}{r_{12}} \phi_i(\mathbf{r}') \phi_j(\mathbf{r}) d\mathbf{r} d\mathbf{r}' \tag{3.1}$$

$\phi_i(\mathbf{r})$ is the single electron orbital. Among various proposed hybrid functionals, B3LYP and HSE are the most commonly used.

In B3LYP (Becke, three-parameter, Lee–Yang–Parr), the exchange-correlation functional is written as

$$E_{XC}^{B3LYP} = E_X^{LDA} + a_0(E_X^{HF} - E_X^{LDA}) + a_X(E_X^{GGA} - E_X^{LDA}) + E_C^{LDA} + a_C(E_C^{GGA} - E_C^{LDA}) \tag{3.2}$$

The parameters in the above expression are: $a_0 = 0.2$, $a_X = 0.72$ and $a_C = 0.81$. The GGA exchange functional E_X^{GGA} and correlation functional E_C^{GGA} are the ones proposed by Becke and Lee, Yang and Parr respectively. The LDA correlational functional is from Vosko, Wilk and Nusair.

HSE (Heyd–Scuseria–Ernzerhof) method offers a range-separated hybrid functional where the exchange part of the Coulomb potential between electrons is decomposed into short and long ranged parts. The correlation part is described by PBE. The exchange-correlation functional is expressed as

$$E_{XC}^{HSE} = \frac{1}{4} E_X^{SR}(\mu) + \frac{3}{4} E_X^{PBE,SR}(\mu) + E_X^{PBE,LR}(\mu) + E_C^{PBE} \tag{3.3}$$

In the above equation, SR and LR correspond to 'short range' and 'long range' respectively. μ is the screening parameter that defines the range separation. The typical value used is $0.2–0.3\,\text{Å}^{-1}$. Two versions of HSE are used, viz., HSE03 [6] and HSE06 [7].

The hybrid functionals yield a much better HOMO-LUMO gap for the molecules compared to LDA/GGA. Also, calculated densities of states compare with the photoemission spectra up to certain level of satisfaction. However, the peak positions and corresponding energy widths of the projected orbitals do not always reproduce the experimental features. More sophisticated theories must be used to treat the correlated orbitals residing at the transition metal center in the molecule.

3.3 Electron Correlation

As the 3d metal centers in phthalocyanine molecules have narrow d orbitals, the description of electronic structure is not proper within LDA or GGA due to electron–electron correlation. One of the popular ways to treat this effect is to introduce a modified DFT (LDA or GGA) functional within the Hubbard framework [8, 9]. The total energy functional within LDA+U has the following form.

$$E_{LDA+U} = E_{LDA}[\rho(\mathbf{r})] + E_U[\{n_{mm'}^{I\sigma}\}] - E_{dc}[\{n^{I\sigma}\}], \qquad (3.4)$$

where E_U is the Hubbard term taking into account the electron correlation and E_{dc} is the double counting term to eliminate the non-classical electron–electron interaction already taken into account in the exchange-correlation term included in LDA. The double counting term is used mainly in two ways, (i) around mean field, (ii) fully localized limit. $n_{mm'}^{I\sigma}$ is the occupation matrix for correlated orbitals m and m' in the spin channel σ residing at the metal atom I. This matrix is obtained by projecting Kohn–Sham orbitals ($\psi_{\mathbf{k}\nu}^\sigma$) on localized orbitals (ϕ_m^I) as

$$n_{mm'}^{I\sigma} = \sum_{\mathbf{k},\nu} f_{\mathbf{k}\nu}^\sigma \langle \psi_{\mathbf{k}\nu}^\sigma | \phi_{m'}^I \rangle \langle \phi_m^I | \psi_{\mathbf{k}\nu}^\sigma \rangle, \qquad (3.5)$$

where $f_{\mathbf{k}\nu}^\sigma$ is the Fermi–Dirac occupation of Kohn–Sham state with band index ν and k-vector $\mathbf{k}$.

Very recently, a few papers [10] have reported results on the application of GW method for phthalocyanine molecules, where G is the single particle Green function and W is the screened Coulomb interaction. The GW method is based on many body perturbation theory to compute quasiparticle energies directly by following Hedin's proposition of writing the self energy as a product of G and W. However, it is not clear whether a perturbative (perturbation on DFT single particle eigenvalues) method or a fully self-consistent one will provide the best result apart from the method's prohibitive computational costs. One should mention that different starting approximations, viz., LDA/GGA, hybrid functionals etc. have been employed to have the best comparison with experimental photoemission spectra. However, more research is needed in this direction as no unanimous solution exists.

3.4 van der Waals Interaction

One of the most important issues of molecule-substrate interactions is the description of weak dispersion interactions. DFT can't describe the weak dispersion interactions. One famous failure is the interplanar separation of graphite using GGA, which produces a much larger value compared to the experimental one. It has been clear demonstrated that these interactions modify the separation between molecules and surfaces and hence modify the electronic structure quite significantly. A recent paper on CoPc/Fe has demonstrated the importance of using this correction to have a good comparison with experimental spin-polarised STM images [11]. These corrections, e.g., van der Waals (vdW) interactions have been implemented in DFT codes in two different ways. The simplified DFT-D method contains pair interactions between atoms characterised by certain physical coefficients. The energy term responsible for dispersion correction reads

$$E_{disp} = \sum_{i=1}^{N_a-1} \sum_{j=i+1}^{N_a} \frac{C_6^{ij}}{R_{ij}^6} f_d(s_6, R_{ij}, R_{6i}, R_{6j}) \qquad (3.6)$$

In the above expression, the coefficient C_6 is species dependent and is possible to calculate from adiabatic connection fluctuation dissipation (ACFD) theorem. The damping function f_d depends on a global scaling factor s_6 and vdW radii of ith and jth atoms along with the separation R_{ij} between them

Grimme's semi-empirical DFT-D2 method is used widely to study the adsorption characteristics of molecules on surfaces. Here the C_6 coefficients are determined from the geometric mean of the free atoms in question. A more rigorous DFT-D method has been proposed by Tkatchenko and Scheffler, where electron densities are used to determine the coefficients.

Apart from DFT-D method, the sparse density functionals, named as vdW-DF have been developed based on ACFD theorem to include nonlocal correlation energy contributions. The nonlocal correlation energy is expressed as

$$E_c^{nl} = \frac{1}{2} \int d^3r \int d^3r' \rho(\mathbf{r}) \phi(\mathbf{r}, \mathbf{r}') n(\mathbf{r}'), \qquad (3.7)$$

where the kernel ϕ takes care of the dispersion corrections. Unlike the pair-interaction type methods, vdW-DFs take into account density-density interactions which fall within the realm of DFT. A number of these sparse DFs have been proposed depending on the type of DFT exchange, e.g., revPBE, rPW86, optB88, optB86b etc. These calculations can be done in two different ways. The relatively cheaper option is to perform non-self consistent calculations by using the self-consistent DFT potentials in a post-processing manner. The more rigorous fully self-consistent calculations are computationally demanding and are rather limited to smaller system sizes.

3.5 Magneto-Crystalline Anisotropy

The information on the easy axis of magnetization of the molecules is important from the point of view of the magnetic coupling between the molecules and magnetic substrates. In this regard, magnetocrystalline anisotropy energy (MAE) is the central quantity of interest. The magnetic molecules may possess large MAE and orbital moments due to their low dimensions and complex arrangements of correlated orbitals. MAE originates from the coupling between lattice and spin. The spin-orbit coupling (SOC) Hamiltonian can be expressed as:

$$H_{SO} = \xi(r)\mathbf{L}.\mathbf{S} = \xi(r)(L_x S_x + L_y S_y + L_z S_z). \qquad (3.8)$$

ξ is the spin orbit coupling constant defined as $\xi = \langle \xi(r) \rangle = \int R_{3d}^2(r)\xi(r)r^2 dr$ where $R_{3d}(r)$ is the radial part of the 3d wave function. MAE can be calculated from DFT in two different ways, (i) using the SOC Hamiltonian as a perturbation to the scalar relativistic Kohn–Sham Hamiltonian, (ii) using a four-component Dirac formalism. In almost all the cases, the two methods yield very similar results.

The second order perturbation theory can also be used to estimate MAE from the following expression.

$$E_{SO} = -\xi^2 \sum_{u,o} \frac{[\langle u \mathbf{L.S} \rangle o \langle o \mathbf{L.S} \rangle u]}{E_u - E_o} \tag{3.9}$$

Here $\rangle o$ and $\rangle u$ correspond to occupied and unoccupied states weighted by the occupations of the d orbitals. $\rangle u/o = \rangle lm$, σ, L and S denote orbital and spin operators and E_u and E_o denote eigenvalues of unoccupied and occupied states, which are obtained from ab initio calculations. It is evident from the expression that MAE increases with the decrease in the denominator containing the difference in eigenvalues. The relative arrangement of Fe-d orbitals is thus very important. A plus (minus) sign of MAE corresponds to an in-plane (out-of-plane) easy axis of magnetization.

3.6 Spin Dipole Moment

One of the most relevant properties for magnetic nanosystems is the magnetic dipole contribution $\langle T_z \rangle$ as it can have a very large value for low dimensional structures, as reported earlier for clusters [12] and organometallics [13, 14]. This has a significant contribution in the effective spin moment, $S_{eff} = M_s + 7\langle T_z \rangle$, which is measured in XMCD experiments, M_s being the saturation spin moment. It should be noted that the sign of $\langle T_z \rangle$ can even be opposite to the spin moment giving rise to a small value of the effective spin moment [13]. The spin dipole contribution can be calculated using the following formalism prescribed in the literature [15–17].

The spin dipole operator can be defined as (Oguchi et al.) [17]

$$T = \sum_i Q^{(i)} s^{(i)}, \tag{3.10}$$

where, $Q^{(i)}$ is the quadrupolar tensor and can be described as:

$$Q_{\alpha\beta}^{(i)} = \delta_{\alpha\beta} - 3\hat{r}_\alpha^{(i)}\hat{r}_\beta^{(i)} \tag{3.11}$$

Every component of T can be written in second quantization form as

$$T_\pm = T_x \pm iT_y = \sum_{\gamma\gamma'} T^\pm_{\gamma\gamma'} a^\dagger_\gamma a_{\gamma'} \quad T_z = \sum_{\gamma\gamma'} T^z_{\gamma\gamma'} a^\dagger_\gamma a_{\gamma'} \tag{3.12}$$

The matrix elements of $T_\pm$ and T_z are:

$$T^\pm_{\gamma\gamma'} = \langle\gamma| c_0^2 s_\pm - \sqrt{6}c_{\pm2}^2 s_\mp \pm \sqrt{6}c_{\pm1}^2 s_z |\gamma'\rangle \tag{3.13}$$

$$T^z_{\gamma\gamma'} = \langle\gamma| -\sqrt{\frac{3}{2}} c_{-1}^2 s_+ + \sqrt{\frac{3}{2}} c_1^2 s_- - 2c_0^2 s_z |\gamma'\rangle \tag{3.14}$$

$|\gamma\rangle = |lm, \sigma\rangle$.

$\langle T_z \rangle$ can easily be calculated from the knowledge of the density matrix obtained from DFT. Even the angular dependence of $\langle T_z \rangle$ observed in experiments [14] can be theoretically reproduced [18].

References

1. Hohenberg P, Kohn W (1964) Phys Rev 136:B864
2. Kohn W, Sham LJ (1965) Phys Rev 140A:1133
3. Jones RO, Gunnarsson O (1989) Rev Mod Phys 61:689
4. Burke K (2012) J Chem Phys 136:150901
5. Jones RO (2015) Rev Mod Phys 87:897
6. Heyd J, Scuseria G, Ernzerhof M (2003) J Chem Phys 118:8207
7. Heyd J, Scuseria G, Ernzerhof M (2006) J Chem Phys 124:219906
8. Anisimov V, Aryasetiawan F, Lichtenstein A (1997) J Phys: Condens Matter 9:767
9. Cococcioni M, de Gironcoli S (2005) Phys Rev B 71:035105
10. Marom N, Ren X, Moussa JE, Chelikowsky JR, Kronik L (2011) Phys Rev B 84:195143
11. Brede J, Atodiresei N, Kuck S, Lazic P, Caciuc V, Morikawa Y, Hoffmann G, Bluegel S, Wiesendanger R (2010) Phys Rev Lett 105:047204
12. Sipr O, Minar J, Ebert H (2009) Eur Phys Lett 87:67007
13. Bhandary S, Ghosh S, Herper H, Wende H, Eriksson O, Sanyal B (2011) Phys Rev Lett 107:257202
14. Stepanow S, Mugarza A, Ceballos G, Moras P, Cezar JC, Carbone C, Gambardella P (2010) Phys Rev B 82:014405
15. Wu R, Freeman A (1994) Phys Rev Lett 73:1994
16. Stöhr J, König H (1995) Phys Rev Lett 75(20):3748
17. Oguchi T, Shishidou T (2004) Phys Rev B 70(2):024412
18. Herper HC, Bernien M, Bhandary S, Hermanns CF, Krüger A, Miguel J, Weis C, Schmitz-Antoniak C, Krumme B, Bovenschen D, Tieg C, Sanyal B, Weschke E, Czekelius C, Kuch W, Wende H, Eriksson O (2013) Phys Rev B 87:174425

Chapter 4
Electronic Structure of Isolated Molecules

Abstract Phthalocyanine molecules with a 3d transition metal in the center, like MnPc, FePc, CoPc, NiPc and CuPc, have attracted a huge interest in the last decades due to the large number of possible applications. Experimental and theoretical gas phase studies are an important reference to understand the properties of the molecules, as well as how they can be modified and manipulated upon deposition on substrates or in supramolecular conformations. However, in several 3d metal phthalocyanines the electronic structure of the single molecule is still under debate even after several spectroscopical studies and computational works have been performed. This is mostly due to the highly correlated 3d electrons of the metal atoms, which pose a challenge for the theory. In addition, the experiments to determine the electronic structure are often carried out in different conditions (on thick films or in gas phase for example), and this can lead to different results. The following chapter provides an overview of the theoretical and experimental results and debates related to the electronic structure of gas phase MnPc, FePc, CoPc, NiPc and CuPc.

4.1 Phthalocyanines with Transition Metal Center

In the present review, the 3d transition metal phthalocyanines MnPc, FePc, CoPc, NiPc and CuPc will be considered, where the metal ion hosted in the center has a formal charge of +2. A full description of the valence band electronic structure is crucial for the study of the alignment of the energy levels in organic semiconductors, as well as a prerequisite to understand and manipulate the spin state in single molecule magnets. The purpose is to unfold the occupations and energies of the frontier orbitals, and especially to distinguish those with metallic character with respect to the other π and σ orbitals localized on the organic shell. In this sense, PES and NEXAFS spectroscopies are particularly powerful, and they have been widely applied to the study of TMPc.

The examination of the molecules in gas phase is fundamental to determine how the molecular properties are affected by different types of intermolecular interactions like by the adsorption on surfaces, or when part of supermolecular architectures, which are relevant for most practical devices. These studies give also an insight into the interaction of the central metal with the organic part of the molecule. The strong ligand field exerted by the ring, splits the 3d orbitals on the metal ion. Assuming the plane of the molecule as laying in the xy surface, and the D_{4h} symmetry group, the resulting d levels are the d_{xy} (with symmetry b_{2g}), two degenerate d_{xz} and d_{yz}, usually indicated as d_π (with symmetry e_g), d_{z^2} (with symmetry a_{1g}) and $d_{x^2-y^2}$ (with symmetry b_{1g}). We will keep this notation throughout the present discussion, but it has to be observed that the molecular symmetry can be reduced to a D_{2h} by Jahn Teller distortions or by intermolecular interactions, like the adsorption on surfaces. The occupation of these orbitals depends on the specific metal ion, and the energy alignment depends on the interaction with the organic shell.

As will be discussed in the following sections, theoretical and experimental studies have often lead to different and contrasting conclusions, which could indicate either the possibility of coexistence of different electronic states or that different kinds of samples have different electronic states. A reason for the controversial results could be the open 3d shell. The strongly correlated electrons can generate several electronic states with close-lying energies and with different spins. As a consequence, no consensus has been achieved so far on the ground state electronic structure of these molecules, either isolated or in crystalline form.

Only few experiments have been performed so far on gas phase samples [1–3]. Most often, for the determination of the electronic structure, samples like thick multilayer films are used, where, however, solid-state interactions of electrostatic and magnetic nature may be relevant. Several works have focused on valence band UPS and several of these have exploited the elemental selectivity by varying the photon energy. This approach has been used for different metal Pc's [1–6]. It is a common practice in UPS studies to compare the measured photoelectron spectra of the valence band (VB) of TMPcs to the theoretical DOS obtained from DFT calculations according to the Koopman's Theorem, [1–3, 7]. In many cases, the Kohn-Sham (KS) ground state eigenvalues of the VB, and especially those obtained with hybrid functionals, have shown to be a good representation of the electron binding energies, and are successfully used to interpret the UPS spectra. To facilitate the comparison with the measurements, the KS eigenvalues are usually convoluted by Lorentzian or Gaussian functions that mimic the spectral broadening due to life time and to experimental factors respectively.

It emerges from all these studies that the most notable differences between the VB electronic structure of the 3d TMPcs are in the molecular orbitals closest to the energy gap, as dictated by the different occupancy of the 3d states of the metal ion. In general a detailed theoretical description is necessary to interpret the experimental data. Due to the strong correlation which characterises the 3d electrons in the transition metals, the DFT outcomes are expected to be considerably influenced by the choice of the DFT functional. A critical analysis of the DFT (with Gaussian basis sets) results for several TMPc molecules (CuPc [8], CoPc and NiPc [9], FePc and MnPc [10]) was carried out by Marom and Kronik in recent years. In these works, several GGA and hybrid functionals were tested, namely VWN, PBE, B3LYP, and HSE for CuPc and PBE, B3LYP, PBEh and M06 for the other TMPc's. Depending on the exchange/correlation functional and basis set chosen, TMPc's have shown to relax into more than one electronic configuration, often very close in energy to each other (for example in the order of less than 0.1 eV in MnPc), and with a different energy sequence of the electronic orbitals. In most cases, these electronic configurations, and especially those obtained with hybrid functionals, give satisfactory agreement with the available UPS measurements. The drawback is that the good matching hampers the possibility to unambiguously determine the electronic structure of the molecule. The GGA approach is affected by the self-interaction error (SIE) and results in an overall contraction of the order of 30% of the energy scale with respect to the UP spectra. Reference [9] In general the LDA/GGA functionals provide different orbital energies and orbital ordering, resulting in a in worse agreement with the experiments. In PW approaches, the addition to the 3d metal states of a Hubbard U term of the magnitude of a few eV has shown to be necessary to correct the energy distribution of the 3d orbitals, reaching an improved agreement with the photoelectron spectra [1, 3, 11].

Recently, theoretical works on TMPc's based on the GW approach have been performed, like on CuPc [12], ZnPc [13] and CoPc [14]. The main hinder to using the GW method is the higher computational cost with respect to the DFT methods. Calculations of larger molecules are usually not employing the fully self consistent GW, but rather the *one shot* G0W0. References [12, 14] In all these cases, the GW DOS had achieved an improved energy alignment with the experimental energies.

In spite of these contradicting experimental and theoretical reports, several comparative studies have been published, for example in Grobosh et al. [15], as part of a series of works based on experimental spectroscopy and DFT calculations. References [6, 16] By studying the VB population of these molecules in form of films adsorbed on metal surfaces, and in particular the HOMO level, the authors report that the energy of the 3d orbitals of the phthalocyanine metal center are closer to the HOMO in Mn-, Co- and FePc than in Cu- and NiPc. Due to this the 3d electrons have a stronger impact on the electronic properties of the former molecules, i.e. Mn-, Co- and FePc.

In the following paragraphs we will briefly summarise the main results obtained in recent years about the ground state electronic structure of isolated 3d TMPc in the gas phase from theoretical and experimental studies.

4.2 MnPc

MnPc has attracted considerable attention as a single molecule magnet candidate since the magnetic properties of single crystal films of MnPc were first measured four decades ago. Reference [17] Interestingly, by means of FLAPW calculations, Kitaoka et al. [18] and Wang et al. [19] have found an out of plane magnetic anisotropy for MnPc, amounting to about 3 meV. In addition, Kondo resonance of MnPc deposited on a Pb surface was studied and manipulated [20]. Various electronic configurations which have been proposed so far for the isolated MnPc are summarised in Table 4.1, alongside the theoretical or experimental methods used in each work. In the Mn atom, the 3d shell is half occupied, with three possible spin states, namely low spin ($S = 1/2$), intermediate spin ($S = 3/2$) and high spin ($S = 5/2$). All the reviewed DFT calculations performed on MnPc with different functionals, as well as the experimental studies, agree on the $S = 3/2$ spin state for the single isolated molecule. The experimental works suggest in some cases the coexistence of more than one molecular electronic ground state (GS) [21], or the possibility of having different GS in different kinds of samples [3]. A $S = 3/2$ spin state and the $^4A_{2g}$ electronic GS for MnPc was obtained in 1970 by Barraclough et al. [21], by measurements of the average magnetic susceptibility (χ) and of the magnetic anisotropy of a single crystal MnPc. In the same paper it was also proposed that the MnPc sample could be in a mixed state of $^4A_{2g}$ and ^4E_g. Reference [21] Several Mn L-edge NEXAFS experiments have been carried out during the last two decades, as shown in Table 4.1. Measurements performed either on a β-MnPc sample [22, 23] or on MnPc films deposited on Ag and Au substrates [24–26] have found evidence for a ^4E_g GS. However in the study by Kroll et al. [26] it is reported that not only a ^4E_g but also a 3d occupation like $(d_{xy})^1 (d_\pi)^2 (d_{z^2})^2$ was compatible with the experimental data.

The majority of the DFT studies have resulted either in a GS configuration with ^4E_g symmetry [3, 19, 28, 30–32] or $^4A_{1g}$[9, 10, 29], or $^4A_{2g}$ [9, 10], see Table 4.1. Liao et al. [32] and Brumboiu et al. [3] present in addition the total energy for some excited electronic states. In particular an excited state with $^4A_{2g}$ symmetry is reported in both works, at 0.16 eV [32] and at 0.11 eV [3] higher energy than the respective ^4E_g GS energies.

Marom and Kronik [10] find for all the tested functionals either a $^4A_{1g}$ or a $^4A_{2g}$ structure, with HOMO and HOMO-1 orbitals having b_{2g} and a_{1u} or a_{1g} and a_{1u} symmetry respectively. A $^4A_{1g}$ symmetry term was obtained also by Wu et al. [29], with a a_{1u} HOMO and a b_{2g} HOMO-1. Stradi et al. [7] investigated several hybrid and PBE X/C functionals (B3LYP, PBE, PBE0, HSE and HSE06) using both the PW and GTO approaches and performed a comparison of the obtained KS orbitals with UPS results for thick and thin films. They find for all the functionals a LUMO with Mn d_{xy} character, with higher metallic contribution in the hybrid cases.

Arillo et al [33] find metallic character in the HOMO ($d_{x^2-y^2}$) and ligand character in the LUMO. In a few cases, the Mn atoms have been described by adding the Hubbard U term, accounting for the Coulomb interaction on the 3d electrons. An example is the work of Calzolari et al. [34], who have used a Hubbard U of 2 and 3 eV and Grobosch et al. [16].

Several UPS studies which have analysed the VB of MnPc are listed in Table 4.1. Grobosch et al. [6, 15, 16] have shown that in comparison to other TMPc's like FePc and CoPc, the HOMO feature of MnPc in the spectra lies at lower binding energy than in the other 3d TMPcs, and consists of a singly occupied molecular orbital (SOMO) level of d_π character and e_g symmetry. DFT Kohn-Sham eigenvalues of MnPc were compared to measured UPS spectra in the work by Marom and Kronik [10], Stradi et al. [7] and Brumboiu et al. [3]. Stradi et al. [7] find in all cases a a_{1u} HOMO. Brumboiu et al. [3] exploit the effect of the photoionisation cross sections, and show that the single ^4E_g electronic state can represent the UPS spectra of a MnPc

Table 4.1 Summary of recent experimental and theoretical results for the electronic structure of the VB of MnPc. The GS symmetry terms proposed so fare are reported in the upper part, and the studies of the VB and DOS in the lower part

Experimental results

Technique	Type of sample	3d occupation and symmetry term[a]	Reference
χ, MA	β-phase crystal	$^4A_{2g}$, ^4E_g	[21]
LXAS	β-phase crystal	$(d_{xy})^1$ $(d_\pi)^3$ $(d_{z^2})^1$, ^4E_g	[22]
LXAS,RPES	films on Ag(111),Au(100)	$(d_{xy})^1$ $(d_\pi)^3$ $(d_{z^2})^1$	[24, 25]
LXAS	films on Au(001),Au(111)	$(d_{xy})^1$ $(d_\pi)^3$ $(d_{z^2})^1$ and $(d_{xy})^1$ $(d_\pi)^2$ $(d_{z^2})^2$	[26]
LXAS, XMCD	polycrystalline β-MnPc	$(d_{xy})^1$ $(d_\pi)^3$ $(d_{z^2})^1$, ^4E_g	[23]

Theoretical results

DFT package	Exc functional	3d occupation and symmetry term[a]	Reference
ADF	VWN+B, Perdew	$(d_{xy})^1$ $(d_\pi)^3$ $(d_{z^2})^1$, ^4E_g	[27]
G09	B3LYP	$(d_{xy})^1$ $(d_\pi)^3$ $(d_{z^2})^1$, ^4E_g	[3]
SIESTA	PBE	$(d_{xy})^1$ $(d_\pi)^3$ $(d_{z^2})^1$	[28]
DFT FLAPW	PBE	$(d_{xy})^1$ $(d_\pi)^3$ $(d_{z^2})^1$, ^4E_g	[19]
G03	PBE, PBEh	$(d_{xy})^2$ $(d_\pi)^2$ $(d_{z^2})^1$, $^4A_{1g}$	[10]
	B3LYP, M06	or $^4A_{2g}$	
	B3LYP	$(d_{xy})^2$ $(d_\pi)^2$ $(d_{z^2})^1$, $^4A_{1g}$	[29]
NRLMOL	PBE	$(d_{xy})^1$ $(d_\pi)^3$ $(d_{z^2})^1$, ^4E_g	[30]
VASP		$(d_{xy})^1$ $(d_\pi)^3$ $(d_{z^2})^1$, ^4E_g	[31]
FLAPW		$(d_{xy})^1$ $(d_\pi)^3$ $(d_{z^2})^1$, ^4E_g	[18]

[a]*when given*
χ susceptibility measurements
MA magnetic anisotropy measurements
LXAS Mn L-edge NEXAFS
RPES Resonant PES.

thick film grown on Au(111), while the description of MnPc molecules in gas phase is better obtained by a mixture of at least two states like ^4E_g and $^4A_{2g}$.

4.3 FePc

An intermediate spin S = 1 is accepted for Fe(II)Pc in gas phase. As for MnPc, the many experimental and theoretical studies have not converged onto a unique molecular ground state, as can be seen from the summary of electronic configurations proposed over the years in Table 4.2.

Table 4.2 Summary of recent experimental and theoretical results for the electronic structure of the VB of FePc. The GS symmetry terms proposed so fare are reported in the upper part, and the studies of the VB and DOS in the lower part

Experimental results

Technique	Type of sample	3d occupation and symmetry term[a]	Reference
χ,MA		$(d_{xy})^1 (d_\pi)^4 (d_{z^2})^1\ ^3B_{2g}$	[21]
MA		$(d_{xy})^2 (d_\pi)^3 (d_{z^2})^1\ ^3E_g A$	[35]
MA		$(d_{xy})^2 (d_\pi)^3 (d_{z^2})^1\ ^3E_g A$	[36]
XMCD		^3E_g	[37]
Mössbauer spectr.		$(d_{xy})^2 (d_\pi)^3 (d_{z^2})^1\ ^3E_g A$	[38]
XMCD	β-phase crystal	$(d_{xy})^2 (d_\pi)^3 (d_{z^2})^1\ ^3E_g A$	[39]
LXAS	films on Au(001),Au(111)	$(d_{xy})^2 (d_\pi)^3 (d_{z^2})^1$	[26]

Theoretical results

Package	Exc functional	3d occupation and symmetry term[a]	Reference
ADF	VWN+B, Perdew	$(d_{xy})^2 (d_\pi)^2 (d_{z^2})^2\ ^3A_{2g}$	[27]
FLAPW		^3E_g	[40]
SIESTA		$(d_{xy})^2 (d_\pi)^3 (d_{z^2})^1\ ^3E_g$	[41]
G03	PBE, PBEh,	$^3B_{2g}$	[10]
	B3LYP, M06	or $^3A_{1g}$	
G03	B3LYP	$(d_{xy})^2 (d_\pi)^2 (d_{z^2})^2\ ^3A_{1g}$	[42]
G03	B3LYP	$(d_{xy})^1 (d_\pi)^4 (d_{z^2})^1\ ^3B_{2g}$	[1]
DFT PW (?)	GGA	$(d_{xy})^2 (d_\pi)^3 (d_{z^2})^1$	[19]

[a]*when given*
χ susceptibility measurements
MA magnetic anisotropy measurements
LXAS Mn L-edge NEXAFS
RPES Resonant PES.

A $^3B_{2g}$ ground term with a $(d_\pi)^4 (d_{xy})^1 (d_{z^2})^1$ occupancy of the 3d states was obtained by means of magnetic anisotropy and χ measurements [21], while the $^3E_g(A)$ with a $(d_\pi)^3 (d_{xy})^2 (d_{z^2})^1$ structure was achieved by means of Mössbauer spectroscopy [35], and by X-ray diffraction [36, 38]. The $^3A_{2g}$ state, which corresponds to a full occupation of the d_{xy} and of the d_{z^2} and to a half occupation of the d_π states $((d_{xy})^2 (d_{z^2})^2 (d_\pi)^2)$ was also proposed by means of XMCD [37]. However, these experiments were performed on FePc bis-adducts, in which case the pseudo octahedral ligand field could have some influence on the electronic configuration. More recently angular resolved XAS Fe L edge spectroscopy measurements [26] resulted compatible with two possibilities, either the $^3E_g(A)$ or the $^3A_{2g}$.

Also for this molecule, the theoretical DFT calculations unfold a complex landscape. In particular, Liao and Scheiner [27] by using the ADF program [43] describe three electronic states with close total energies: the $^3A_{2g}$ ground state, a ^3E_g state at

0.07 eV at higher energy and a $^3B_{2g}$ at 0.1 eV higher energy. More recently Sumimoto et al. [42] in B3LYP/DFT calculations obtained a $(d_{z^2})^2 (d_{xy})^2 (d_\pi)^2$ GS configuration. Marom and Kronik [10] report either a $^3B_{2g}$ GS, with a a_{1g} HOMO and e_g HOMO-1 for the hybrid functionals, inverted in the case of the GGA functionals, or a $^3A_{1g}$, with a_{1u} HOMO and a_{1g} HOMO-1, in agreement with Sumimoto et al. [42]. The ^3E state resulted also from charge transfer multiplet model calculations [44] to interpret a Fe L-edge NEXAFS of FePc powder [45], Gas phase UPS exploiting the Pcs, in comparison with DFT calculations found the lack of metal character of the HOMO in Ref. [1], by varying the photon energies from 20 to 130 eV and in Ref. [2] by using HeI and HeII radiation.

4.4 CoPc

CoPc is a low spin molecule (S = 1/2) [46], with three holes available in the 3d shell. Assuming an empty $d_{(x^2-y^2)}$, the last hole can be located in one of the remaining d states, namely d_{z^2}, d_{xy} or d_π, allowing for different possible configurations. Several experimental and theoretical works have identified an unpaired electron in the d_{z^2} orbital [26, 33, 47–50]. Two electronic configurations for the GS have been proposed: the $^2A_{1g}$ by means of unrestricted Hartree Fock Slater (UHFS) discrete variational (DV) $X\alpha$ calculations [48] and DFT [19, 51], and the ^2E_g by means of DFT. Reference [27] In the latter, two higher energy electronic states were computed: the $^2A_{1g}$ at 0.11 eV from the GS and the $^2A_{1u}$ at 0.31 eV from the GS. Reference [27] Some UPS studies on CoPc films [2, 6, 52, 53] support the presence of a strong metallic component in the HOMO. Even for this molecule the theoretical DFT description has not provided a uniform picture. The metallic HOMO of d_π character is supported by Liao and Scheiner [54], and by Białek et al. [49] who obtain a metallic HOMO (a d_{z^2} or e_g) and a ligand LUMO. Marom and Kronik [9] report an a_{1u} HOMO for PBE, B3LYP, PBEh and M06, which is a π orbital localized on the aromatic ring and an e_g orbital with d_π contribution as the LUMO. Arillo et al. [33] find that both HOMO and LUMO have ligand character (a_{1g} and $2e_g$ respectively).

4.5 NiPc

NiPc is a closed shell compound, with a 3d^8 configuration for the Ni atom. In contrast to the other molecules discussed above, the electronic structure of NiPc is considered less controversial, and has not generated the same flourishing of investigations. However also in this case the electronic structure is not fully understood. A HOMO of a_{1u} character, and a b_{1g} LUMO, formed by the empty $d_{x^2-y^2}$ were suggested already by Discrete Variational $X-\alpha$ (DV $X-\alpha$) calculations in 1986. Reference [55] The same a_{1u} character for the HOMO with a b_{1g} or b_{2g} and with a $^1A_{1g}$ GS symmetry was obtained by the successive DFT studies [27, 51]. However a study based on first principles all-electron full-potential linearized augmented plane wave (FLAPW) energy band method by Bialek et al. [56] found an e_g HOMO, with contributions from the d_{xz} and d_{yz}, and a a_{1u} LUMO, with a contribution of the $d_{x^2-y^2}$ orbitals in the occupied DOS. Differences in orbital occupancy due to the underlying DFT functional have been described by Marom and Kronik [9] find a a_{1u} HOMO and a e_g LUMO.

4.6 CuPc

In CuPc the central metal has a $3d^9$ configuration: the general consensus is that the singly occupied $d_{x^2-y^2}$ forms a molecular orbital of b_{1g} symmetry (the SOMO) hybridizing with the 2p of the nearby N atoms, while the doubly occupied HOMO, a π orbital with a_{1u} symmetry, is located on the organic rings and has mainly contributions from the C 2p. However, the energy ordering and character of the frontier orbitals of CuPc is still under debate. Recent theoretical DFT results have reported different possible orbital energy alignments, with either the SOMO or the HOMO at higher energy. Gas phase UPS measurements of the VB of CuPc have been performed recently by Evangelista et al. [57], and by Vogel et al. [58] and both works are accompanied by B3LYP/DFT calculations. Evangelista et al. [57] find that the b_{1g} SOMO is located at 1.5 eV higher energy than the a_{1u} HOMO. Analogous results had been obtained by Liao and Scheiner [54]. However, B3LYP/DFT calculations by other groups give the opposite picture, with the SOMO lower in energy than the HOMO [16, 29, 58, 59]. In particular, in Ref. [58] the results of a metallic SOMO at about 1 eV lower energy than the HOMO, are also supported by UPS with varying photon energy.

Marom et al. [8] compared LDA, PBE, B3LYP and HSE and report orbital switching between the different approaches. The half unoccupied b_{1g} is the HOMO in LDA, and HOMO-1 in PBE, and the half unoccupied b_{1g} is the LUMO, due to the small separation of less than 1 eV, between the two. Conversely, they find that an a_{1u} orbital is the HOMO and an e_g is the LUMO when using the hybrid B3LYP and HSE functionals, where the separation between the b_{1g} spin up and spin down orbitals is much larger, in the order of a few eV. A different ordering of HOMO and HOMO-1 between LDA and PBE is reported also by Calzolari et al. [34] with the PW_{SCF} code, although they results are not the same as in Ref. [8]. In addition, by applying the PBE+U method with U equal to 3 eV, the ordering of the molecular orbitals is the same as in the LDA. Shen et al. [60], who compare the molecular orbitals given by a PBE/DFT calculation with the Siesta code with PBE and PBEh performed by G03, report that the level ordering for CuPc with ring based a_{1u} HOMO as the outer molecular orbital is conserved in the three approaches. Further investigations of the electronic structure have been performed by Marom et al. [12] by means of G0W0 calculations based on semilocal (PBE) and hybrid (PBEh) DFT functionals, with FHI-aims and including relativistic effects via the scaled ZORA method. They argue that Although PBEh is a better starting point for the GW calculations when compared to the spectroscopy results, in both cases the GW calculations provide a HOMO at the higher energy, and a SOMO at lower energy. The SOMO is also identified as the source of the peak F, that has eludeded the calculations with the hybrid functionals [12].

References

1. Brena B, Puglia C, de Simone M, Coreno M, Tarafder K, Feyer V, Banerjee R, Göthelid E, Sanyal B, Oppeneer P, Eriksson O (2011) J Chem Phys 134:074312
2. Michael Vogel M, Schmitt F, Sauther J, Baumann B, Altenhof A, Lach S, Ziegler C (2011) Anal Bioanal Chem 100: 673–678
3. Brumboiu I, Totani R, de Simone M, Coreno M, Grazioli C, Lozzi L, Herper HC, Sanyal B, Eriksson O, Puglia C, Brena B (2014) J Phys Chem A 118:927
4. Berkowitz J (1979) J Chem Phys 70:2819
5. Ellis TS, Park KT (2006) J Appl Phys 100:093515
6. Grobosch M, Aristov VY, Molodtsova OV, Schmidt C, Doyle B, Nannarone S, Knupfer M (2009) J Phys Chem C 113:13219
7. Stradi D, Díaz C, Martín F, Alcamí M (2011) Theor Chem Acc 128:497–503
8. Marom N, Hod H, Scuseria GE, Kronik L (2008) J Chem Phys 128:164107
9. Marom N, Kronik L (2009) Appl Phys A 95:159
10. Marom N, Kronik L (2009) Appl Phys A 95:165
11. Bhattacharjee S, Brena B, Banerjee R, Wende H, Eriksson O, Sanyal B (2010) Chem Phys 377:96
12. Marom N, Ren X, Moussa JE, Chelikowsky JR, Kronik L (2011) Phys Rev B 84(19):195143. https://doi.org/10.1103/PhysRevB.84.195143
13. Umari P, Fabris S (2012) J Chem Phys 136:174310
14. Salomon E, Amsalem P, Marom N, Vondracek M, Kronik L, Koch N, Angot T (2013) Phys Rev B 87:075407
15. Grobosch M, Smith C, Kraus R, Knupfer M (2010) Org Electron 11
16. Grobosch M, Mahns B, Loose C, Friedrich R, Schmidt C, Kortus J, Knupfer M (2011) Chem Phys Lett 505:122–125
17. Mitra S, Gregson AK, Hatfield WE, Weller RR (1983) IC 22:1729–1732
18. Kitaoka Y, Sakai T, Nakamura K, Akiyama T, Ito T (2013) JAP 113:17E130
19. Wang J, Shi Y, Cao J, Wu R (2009) Appl Phys Lett 94:122502
20. Fu YS, Ji SH, Chen X, Ma XC, Wu R, Wang CC, Duan WD, Qiu XH, Sun B, Zhang P, Jia JF, Xue QK (2007) PRL 99:256601
21. Barraclough CG, Martin RL, Mitra S, Sherwood RC (1970) J Chem Phys 53:1638
22. Williamson BE, VanCott TC, Boyle ME, Misener GC, Stillman MJ, Schatz PN (1992) J Am Chem Soc 114:2412
23. Kataoka T, Sakamoto Y, Yamazaki Y, Singh VR, Fujimori A, Takeda Y, Ohkochi T, Fujimori SI, Okane T, Saitoh Y, Yamagami H, Tanaka A (2012) Sol State Commun 152:806
24. (2011) J Phys Chem C 115:21334
25. Petraki F, Peisert H, Aygül U, Latteyer F, Uihlein J, Vollmer A, Chassé T (2012) J Phys Chem C 116:11110
26. (2012) J Chem Phys 137:054306
27. Liao MS, Scheiner S (2001) J Chem Phys 22:9780–9791
28. Shen X, Sun L, Yi Z, Benassi E, Zhang R, Sanvito ZSS, Hou S (2010) Phys Chem Chem Phys 12:10805
29. Wu W, Kerridge A, Harker AH, Fisher AJ (2008) Phys Rev B 77:184403
30. Friedrich R, Hahn T, Kortus J, Fronk M, Haidu F, Salvan G, Zahn DRT, Schlesinger M, Mehring M, Roth F, Mahns B, Knupfer M (2012) J Chem Phys 136:064704

31. Nguyen TQ, Padama AAB, no MCSE, Kasai H (2013) ECS Trans 45:91
32. Liao MS, Watts JD, Huang MJ (2005) INORG 44
33. Arillo-Flores O, Fadlallah MM, Schuster C, Eckern U, Romero AH (2013) Phys Rev B 87:165115
34. Calzolari A, Ferretti A, Nardelli MB (2007) Nanotechnology 18:424013
35. Dale BW, Williams RJP, Johnson CE (1968) Thorp TL J Am Chem Soc 49:3441–3444
36. Coppens P, Li L, Zhu NJ (1983) J Am Chem Soc 105:6173–6174
37. Stillman MJ, Thomson AJ (1974) J Chem Soc Faraday Trans 7(2):790–804
38. Filoti G, Kuz'min MD, Bartolomé J (2006) Phys Rev B 74:134420
39. Bartolomé J, Bartolomé F, García LM, Filoti G, Gredig T, Colesniuc CN, Schuller IK, Cezar JC (2010) Phys Rev B 81:195405
40. Białek B, Kim IG, Lee JI (2003) Surf Sci 526:367–374
41. Kuz'min MD, Hayn R, Oison V (2009) Phys Rev B 79:024413
42. Sumimoto M, Kawashima Y, Hori K, Fujimoto* H (2009) Dalton Trans 29:5737–5746
43. Velde GT, Bickelhaupt FM, Baerends EJ, Guerra CF, Gisbergen SJAV, Snijders JG, Ziegler T (2001) J Comp Chem 22
44. de Groot F, Kotani A (2008) Core level spectroscopy of solids. Advances in Condensed Matter Science, vol 6. CRC Press, Boca Raton
45. Miedema P, Stepanow S, Gambardella P, de Groot F (2009) J Phys: Conf Series 190:012143
46. Assourl JM, Kahn WK (1965) J Am Chem Soc 87:207–212
47. Figgis B, Kucharski ES, Reynolds PA (1989) J Am Chem Soc 111(1683):1692
48. Reynolds PA, Figgis BN (1991) IC 30(10):2294. https://doi.org/10.1021/ic00010a015. http://dx.doi.org/10.1021/ic00010a015
49. Białek B, Kim IG, Lee JI (2006) Thin Solid Films 513:110–113
50. Bhattacharjee S, Brena B, Banerjee R, Wende H, Eriksson O, Sanyal B (2010) Chem Phys 377:96–99
51. Rosa A, Baerends EJ (1994) Inorg Chem 33(584):595
52. Papageorgiou N, Salomon E, Angot T, Giovanelli JMLL, Lay GL (2004) Prog Surf Sci 77:139–170
53. Barlow DE, Scudiero L, Hipps KW (2004) Langmuir 20(11):4413. https://doi.org/10.1021/la035879l. http://dx.doi.org/10.1021/la035879l PMID: 15969147
54. Liao MS, Scheiner S (2001) J Chem Phys 114:9780
55. Kutzler FW, Ellis D (1986) J Chem Phys 84:1033–1038
56. Białek B, Kim IG, Lee JI (2002) Synth Metals 129
57. Evangelista F, Carravetta V, Stefani G, Jansik B, Alagia M, Stranges S, Ruocco A (2007) J Chem Phys 126:124709
58. Vogel M, Schmitt F, Sauther J, Baumann B, Altenhof A, Lach S, Ziegler C (2011) Anal Bioanal Chem 400:673
59. Aristov VY, Molodtsova OV, Maslyuk VV, Vyalikh DV, Zhilin VM, Ossipyan YA, Bredow T, Mertig I, Knupfer M (2008) J Chem Phys 128:034703
60. Shen X, Sun L, Yi Z, Benassi E, Zhang R, Shen Z, Sanvito S, Hou S (2010) Phys Chem Chem Phys 12:10805

Chapter 5
Electron Correlation and Spin Transition

Abstract Theoretical treatment of functional metalorganics is non-trivial for the metal centers with narrow bands (3d; 4d of transition metals or 4f bands of rare-earth metals), featuring a sizeable Coulomb interaction. An interplay between crystal field, spin-orbit coupling and Coulomb interaction expresses the properties of the molecule. Correlated metal centers, immersed in the electron bath of organic ring makes it ideal to treat with Anderson's impurity model. In this chapter, we will focus on the description of electron correlation in functional metalorganics with the aid of density functional theory, combined with a many body approach. For most of the illustrative purposes, we will consider iron porphyrin (FeP) molecule. The chapter will reveal the importance of the treatment of explicit electron correlation in order to accurately identify the spin transition, magnetic anisotropy, Kondo effect etc., which are key ingredients for molecular spintronics and electronics.

5.1 Electron Correlation: DFT++

The metal-organic compounds consist of metal ions or clusters coordinated with an organic framework. An independent electron ansatz fails to describe metal-organic molecules due to their low dimensionality and inherent confinement effect. The metal center, e.g. in metal-porphyrin (Mp) or metal-phthalocyanine (MPc) with a square planar symmetry, exhibits an extremely strong orbital dependent dynamical hybridization with ligands. The narrow 3d-states of transition metal (TM) or 4f states of rare-earth center feature reasonably strong Coulomb repulsion, which combined with crystal field and spin-orbit coupling result in exotic phenomena, such as, Kondo effect, spin transition, large magnetic anisotropy, many-body resonances etc. The potential application field spans over electronics, spintronics, information storage, quantum computing etc. In MPc/Mp (M = TM, RE), description of the features in metal center requires delicate treatment of many body theory allowing low energy

© The Author(s), under exclusive license
to Springer Nature Singapore Pte Ltd. 2020
H. C. Herper et al., *Molecular Nanomagnets*, Nanoscience
and Nanotechnology, https://doi.org/10.1007/978-981-15-3719-6_5

fluctuations. The organic part of the molecule, however, features quite delocalized states spanning almost the whole molecule. The independent particle theory provides accurate description for this segment. For the realistic material modelling, the philosophy of Anderson's impurity model falls in place which describes a single correlated site connected to a *bath* of conduction electrons, described within the density functional theory (DFT).

The key idea relies on the division of the Hilbert-space. The full hamiltonian of the molecular system, H, is comprised of a local impurity hamiltonian, H_{loc} that is coupled to a delocalised bath consisting mostly s, p electrons.

$$H = \underbrace{\sum_{i,j} \varepsilon_{ij}^d d_i^\dagger d_j + \frac{1}{2} \sum_{i,j,k,l} U_{ijlk} d_i^\dagger d_j^\dagger d_k d_l}_{H_{loc}}$$
$$+ \sum_{i\nu} (V_{i\nu} c_\nu^\dagger d_i + H.c.) + \sum_\nu \varepsilon_\nu c_\nu^\dagger c_\nu, \tag{5.1}$$

Here ε_{ij}^d describes the impurity onsite energies while U_{ijlk} represents the local screened Coulomb interaction. $i, j, k, l = (m, \sigma)$ are combined orbital and spin indices. d_i and c_ν are the impurity and the bath degrees of freedom. U_{ijlk} is parametrized by Slater integrals, F^0, F^2 and F^4 for 3d shell systems. The values of the Slater integrals can be calculated with constrained random phase approximation (cRPA) or constrained local density approximation (cLDA) [1]. For transition metal centers in MP or MPc, F^0 ranges from 3 to 5 eV and J $\sim$ 1 eV. In the above equation, the first two terms represent the electrons in the impurity site. The third one consists of a hybridization parameter, $V_{i\nu}$, which gives the amplitude of the electron hopping between the impurity site and the bath. In the fourth term, ε_ν denotes the bath energy levels. The dynamical, that is energy dependent, part of the hybridization function is comprised of the later two parts while the onsite energies for the static part of it are described by the first two terms. A continuous energy dependent hybridization function can be given by:

$$\Delta_{ij}(\omega) = \sum_k \frac{V_{ik} V_{kj}}{\omega + i\delta - \varepsilon_\nu}. \tag{5.2}$$

Within a DFT++ treatment [11], the hybridization function can be obtained from first principles calculations, and hence the above mentioned parameters of the model. In practice, a local Green function, G_{imp} is constructed from DFT and $\Delta_{ij}(\omega)$ is related to it in the following way:

$$G_{imp}^{-1}(\omega) = \omega + i\delta - \varepsilon_d - \Delta(\omega) = \omega + i\delta - \tilde{\Delta}(\omega). \tag{5.3}$$

here, ε_d is the static crystal field and can be extracted from $\tilde{\Delta}(\omega)$ at the limit of $\omega \to \infty$. It has to be noted that the local Green function is to be described in localized basis. In the case of calculation within plane-wave basis, description of localized basis can be done either via projection [2] or constructing Wannier functions [3].

The local impurity problem [10] can be solved with existing solvers, such as Exact Diagonalization (ED) or continuous time quantum Monte Carlo [4]. The results, shown in this chapter, are obtained employing ED solver. The DFT contribution to the electron correlation is extracted with a double counting (DC) term, which for the following discussion has been calculated within different approximations.

5.2 Dynamical Hybridization Function

Let us now elaborate the scenario in details with an example of a molecular system, iron porphyrin (FeP) and iron Phthalocyanine (FePc). In FeP, the single Fe atom is directly co-ordinated with four N atoms in a square cage format, while N-atoms are strongly hybridized with C atoms in the outer pyrrolic ring, see inset of Fig. 5.1. In FePc, the outer organic ring consists of four more N atoms as can be seen from Fig. 1.1. Both the molecules, as a whole, exhibit D_{4h} symmetry.

In the case of FeP, the bath is formed out of the molecular orbitals of pyrrolic N and aromatic C- p orbitals. It has to be noted that N_p orbitals have the major contribution, as those make chemical bonds with Fe but the role of C_p is significant. The Fig. 5.1 shows the calculated orbital (Fe-3d) dependent real (upper panel) and imaginary (lower panel) parts of the hybridization function of FeP molecule. The

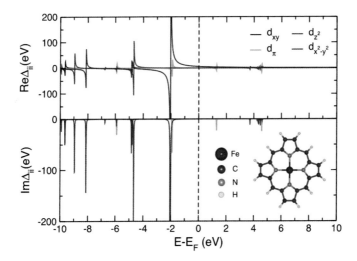

Fig. 5.1 Real and imaginary parts of the hybridization function for Fe in FeP calculated with PBE in the non-spin polarized mode. A smearing parameter of 0.01 eV used for visualization purpose. The geometry of FeP is shown in the inset with the atoms labeled by their types. *Source* The figure is taken from [2] with the permission from American Physical Society

real part $\tilde{\Delta}$ provides the dynamical crystal field while the imaginary part of $\tilde{\Delta}$ quantifies the density of bath states coupling to each impurity orbital weighted by the hybridization matrix elements $V_{i\nu}$. The most dominant peak in $\tilde{\Delta}$ appears at 2.03 eV below fermi energy for the Fe-$d_{x^2-y^2}$ orbital. This arises due to the strong in-plane σ bond formation between $d_{x^2-y^2}$ and axial N-$p_{x/y}$ orbitals. The other in-plane orbital d_{xy} shows almost no hybridization apart from a small peak at 4.8 eV below the Fermi energy. Among the out-of-plane orbitals, d_π (d_{xz}, d_{yz}) orbitals appear closest to the Fermi energy, at -2 eV. Appearance of this peak reflects a $\pi - \pi$ interaction of Fe d_π orbitals with N-p_z orbitals, which is expected in the square planar ligand field of the FeP molecule. The other out-of-plane and in-plane contributions are present in the -4.5 to -10 eV energy range, which play an insignificant role due to the distance from Fermi energy. It can be deduced that the Fe-$d_{x^2-y^2}$ orbital is the most important candidate to feature properties of the molecule owing to it's amplitude and the proximity to the Fermi energy.

Both the coupling amplitude and the bath energy rely on the position of the molecular orbitals formed by the organic ligands. It is, hence, different for different molecules, despite of having same symmetry. FePc exhibit similar structure of the hybridization function [2]. The size of the organic ring is, however, much bigger compared to FeP, resulting in a much stronger bath-site coupling (3.67 eV) for Fe-$d_{x^2-y^2}$ orbital. The most dominant peak appears closer (1.98 eV) to Fermi energy, compared to that of FeP.

5.3 The Spin Transition

The structure of the hybridization function remains mostly the same for porphyrin and phthalocyanine molecules exhibiting D_{4h} symmetry; i.e.the dominant contributor in the dynamical hybridization is $d_{x^2-y^2}$ while the other orbitals play an insignificant role. It has also be seen that [2] the major peak close to the Fermi energy is enough to dictate the molecular properties. However, different TMs in the center and different organic rings result in different sizes of the bath-impurity coupling and position of the bath energies. The static part of the hybridization function, as well, shows a large crystal field separation between $d_{x^2-y^2}$ and the 'rest of the d-orbitals', d_{rest} while crystal field separation within d_{rest} is comparatively small. It is thus sensible to introduce a single crystal field, E_{cryst} separating $d_{x^2-y^2}$ and d_{rest} and single hopping amplitude, V, between bath and $d_{x^2-y^2}$ to model a spin crossover scenario in this class of molecules [2]. In this downfolded model, we will have a simultaneous variation in E_{cryst} and V, which in practice is a result of mechanical strain on the molecule [5]. At this stage, we make an assumption that even though the Fe-N bond length is changed, the rest of the molecular structure remains almost the same, resulting in no change in the bath energy position. Iron in FeP/FePc possesses a 2+ valence state; the resulting possible spin states are the low spin state (S = 0), the intermediate spin-state (S = 1) and the high spin state (S = 2). In a free FeP/FePc molecule, achieving S = 0 state is non-trivial and molecules exhibit $S = 1$ spin state, while exposed in a weaker crystal field, $S = 2$ state is achieved. Hence, discussion of spin-crossover will be around $S = 1$ (which we will consider as low spin state from now on) and $S = 2$ spin states. The origin of the spin-states is an interplay between crystal field, bath-site electron hopping, intra-atomic Hund's exchange and the Coulomb interaction. In a model calculation, we keep U and J fixed while varying E_{cryst} and V continuously to estimate a parameter space for a particular spin-state realization. Figure 5.2 depicts such a spin-phase diagram of FeP molecule described in the parameter-space of E_{cryst} and V. Individual phases are characterised by dominant contributions and demonstrated with RGB colour code. The characteristic energy contribution to the atomic (V → 0) low spin (LS) state is the gain in crystal field energy over Hund's exchange. The dominant contributor (expressed in red) here is $E_{LS}=E_{cryst}(1-n_{x^2-y^2}+(2-n_p))$, where $n_{x^2-y^2}$ and n_p are occupation in $d_{x^2-y^2}$ and the bath, respectively. However LS appears for sufficiently high crystal field; if reduced at a transition value, E_{trans}, the spin state switches to an atomic (V → 0) high spin (HS) state, occupying $d_{x^2-y^2}$ orbital. The characteristic energy contribution (expressed in blue) is of exchange origin; $E_{HS}=I_{ex}S^{d_{x^2-y^2}} \cdot S^{rest}$, where I_{ex} is the Stoner parameter and $S^{d_{x^2-y^2}}$ and S^{rest} are spins in $d_{x^2-y^2}$ and d_{rest}. Once the electron hopping is sufficiently large, the spin state remains S=1 but is governed by bonding (bonding state) rather than crystal field with a characteristic energy term (expressed in green): $E_{BS}= V(d^{\dagger} p + h.c.)+E_p(n_p-2)+(E_d-E_{cryst})(2-n_p)$. It has to be noted that, the E_{trans} with higher V is reduced and the reason is that both bonding and crystal field shift $d_{x^2-y^2}$ orbital in the same direction. At a sufficiently large V but with a significantly low E_{cryst}, a high spin state can be achieved in order to gain exchange energy. The phase is termed as anti-bonding state (AS), and can be identified in the phase diagram where there is sufficient admixture of green and blue. The phase boundary is defined in V and E_{cryst}, which follows a quadratic dependence at the boundary. It has to be mentioned that one needs to be careful about the double counting (DC) term here. All the points in Fig. 5.2 are obtained with a "mid-gap" double counting [2]. However, the uncertainty of DC allows us to vary it but constraining the number of particles in the system. The error-bars correspond to that variation of DC, which essentially provides us a "transition-area" instead of a "transition-line".

The spin-phase diagram gives us a qualitative estimation, one could employ to achieve externally stimulated spin crossover. From a realistic materials calculation for free FeP, one finds a quite strong V (3.36 eV) resulting in a BS low spin state. In practice, however, the molecule can be stretched; e.g.chemisorption [6] and physisorption [5, 9] on surface, or by adsorbing additional ligands [7]. This allows us to perform a "Gedanken experiment" by uniformly stretching the molecule. The estimation of the stretch is given by the Fe-N bond length and as is seen in Fig. 5.2 (orange points), one can push the molecule to the spin-transition regime with a stretching that amounts to a Fe-N bond-length of 2.11Å or more.

The spin-crossover situation in FeP can be more vivid by the estimation of the spin-transition energy, E_{LS}-E_{HS}. In Fig. 5.3 (upper panel) we have shown the spin transition energy for different strain on the molecule. A comparison is also done with different treatments of electron correlation; LDA+U (Dudarev, Lichtenstein),

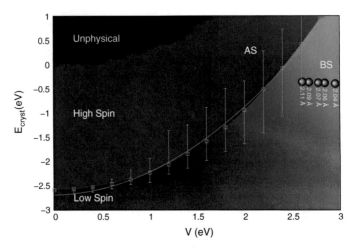

Fig. 5.2 The phase diagram depicting the spin states of FeP with the tuning of the static crystal field E_{cryst} and the hybridization strength V. The phase boundary is accompanied by the allowed values of E_{cryst} for fixed values of V. The blue curve is a result of fitting with a tight-binding model. AS and BS indicate antibonding and bonding regions. Calculated values of E_{cryst} and V from DFT (non spin-polarised PBE) are shown in orange circles along with the corresponding Fe-N bond lengths. *Source* The figure is taken from [2] with the permission from American Physical Society

DFT++ (DC: midgap, FLL). In general, one can observe a spin-transition with in LDA+U for Fe-N bond length around 2.03–2.06 AA but no spin-transition is observed in DFT++, in the considered regime of strain. The uprising value of spin-transition energy, however, suggest a crossover if stretched further. The difference in spin crossover energy form DFT++ and PBE+U approaches is evident from Fig. 5.3. The inclusion of dynamic correlation in DFT++ results in multiple Slater determinant states which can be observed from the occupations corresponding to two different spin states. Almost an integer occupation is observed in high spin state suggesting a maximum weightage over a single Slater determinant. On the contrary, the low spin-state produces fractional occupations, reflecting the presence of multiple Slater determinants. In PBE+U, this symmetry is broken by choosing a single Slater determinant and hence results in an additional energy cost. This can be clearly observed in the upper panel of Fig. 5.3, where PBE+U results much smaller value of $E_{LS} - E_{HS}$ compared to DFT++. This signifies the importance of dynamic electron correlation. The symmetry breaking within PBE+U has a significant consequence in the estimation of magnetic anisotropy. In the lower panel of Fig. 5.3, calculated magnetic anisotropy energy (MAE) within DFT++ (DC: FLL) shown for molecule under strain. The values are at least an order higher compared to that calculated within PBE+U. In all cases, the easy magnetization axis lies in the plane of the molecule, which is also observed experimentally recently [8].

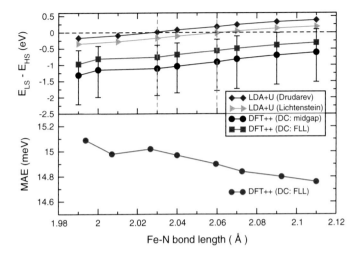

Fig. 5.3 (upper panel) Energy differences between low spin (LS) and high spin (HS) states calculated by DFT++ (double counting (DC) treated by placing chemical potential in the middle of the gap and DC calculated using fully localized limit (FLL)) and PBE+U (double counting by Dudarev and Lichtenstein) methods as a function of Fe-N bond lengths in FeP are shown. The error bars correspond to the variation of onsite energy (model Hamiltonian in Eq. 5.1) in DFT++ calculations. (Lower panel) Magnetic anisotropy energy (MAE) as a function of Fe-N bond lengths calculated with the double counting correction in the fully localized limit. *Source* The figure is taken from [2] with the permission from American Physical Society

5.4 Kondo Effect and Hybrdization Function

Beyond the traditional Kondo systems, e.g. dilute magnetic impurities in metals, magnetic clusters or adatoms on metallic surfaces, interest has grown up significantly in recent years on magnetic organic molecules. MPc mostly with transition metal center has been of particular interest from that aspect. The key point lies on the spin state of the metal center and interaction with the ligands, that provides the conduction electron bath. The spectroscopic signature for the Kondo effect can be observed from photo electron spectroscopy (PES). The Kondo systems of adatoms/clusters or magnetic molecules on substrate can be studied by scanning tunnelling spectroscopy (STS) by combining atomic scale resolution of scanning tunnelling microscopy (STM) with an energy resolution of few μeV. From the theoretical aspect, the single particle ansatz is inadequate because of its pure many body origin. Technically, it is an ideal system for DFT+AIM, methods, such as DFT++.

TMPc or TMP with a single TM at the center are the closest to typical Kondo system of a single adatom on nonmagnetic metallic surface and thus has been explored quite extensively. The generic features of the hybridization function is mostly common in TMPc or TMP but not all of those exhibit Kondo effect. The reason can be identified primarily from the hybridization function and the local correlation effect. The local density of states in a free molecule is gapped and hence it is hard to observe any Kondo effect. TMPcs adsorbed on nonmagnetic surfaces, such as Ag(001), Pb

Fig. 5.4 Hybridization function for 3d TM electrons **a** real part, **b** imaginary part. Reprinted with permission from Nano Lett. 14, 3895 (2014) Copyright (2014) American Chemical Society

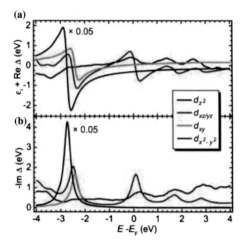

(111) expose the possibility of it; not all TM, however, respond in the same way. In spite of being a typical ferromagnet, Fe, Co centered Pc do not exhibit Kondo resonance while NiPc, MnPc and CuPc show clear Kondo resonance, even though Cu is nonmagnetic.

Figure 5.4 represents real and imaginary part of the hybridization function TMPc on Ag(001) surface. The smooth featureless behaviour near Femi energy suggest the strong hybridization with Ag-s,p orbitals. Among all 3d-orbitals d_{z^2} couple strongly to the substrate underneath but extremely weakly with the ligands. The $d_{xz/yz}$ orbitals couples with ligand out-plane-orbitals of the ligands which interact as well with out-of-plane orbitals from the surface. The appearance of feature at Fermi energy suggest that one could expect Kondo screening for d_{z^2} and $d_{xz/yz}$ orbitals. These features are mostly generic for transition metals. But this by itself is not enough for the Kondo resonance. The orbital population, which is a direct consequence of local Coulomb interaction among the 3d shell, has the conclusive role. An orbital occupation close to half-filling can lead to the Kondo-resonance. The half-filled d_{z^2} appears in CuPc or MnPc, while the orbital filling is almost complete in FePc or CoPc, resulting in diminishing the Kondo resonance. The $d_{xz/yz}$ orbital also show a Kondo peak but in this case the conduction electron comes from the molecular orbital, close to Fermi level that is broadened by the coupling to the substrate. The other two orbitals, $d_{x^2-y^2}$ and d_{xy} do not couple to the substrate near Fermi energy and remain inert from the aspect of the Kondo resonance.

References

1. Scherlis DA, Cococcioni M, Sit P, Marzari N (2007) J Phys Chem B 111:7384
2. Bhandary S, Schüler M, Thunström P, di Marco I, Brena B, Eriksson O, Wehling T, Sanyal B (2016) Phys Rev B 93:155158

3. Kügel et al (2014) Nano Lett 14(7):3895–3902
4. Gull et al (2011) Rev Mod Phys 83:349
5. Bhandary S et al (2011) Phys Rev Lett 107:257202
6. Bhandary S et al (2013) Phys Rev B 88:024401
7. Wackerlin C, Chylarecka D, Kleibert A, Müller K, Iacovita C, Nolting F, Jung TA, Ballav N (2010) Nat Commun 1:1057
8. Heinrich BW, Braun L, Pascual JI, Franke KJ (2015) Nano Lett 15:4024
9. Bhandary S, Eriksson O, Sanyal B (2013) Nat Sci Rep 3:3405
10. Anderson PW (1961) Phys Rev 124:41
11. Karolak M, Wehling TO, Lechermann F, Lichtenstein AI (2011) J Phys Condens Matter 23:085601

Chapter 6
Interaction with Substrates

Abstract Organometallic molecules have attracted interest because their properties can be varied by changing ligands, metal center, end groups etc. which makes them candidates for various applications. Special attention has been paid to hybrid systems of molecules and substrates as possible building blocks for future electronic and magnetic devices. In view of such devices phthalocyanine molecules are advantageous because they can adsorb flat on metallic or semiconducting substrates. Aiming to understand the magnetic properties of the molecules and their interplay with substrates and ligands the focus will be on the paramagnetic Pc molecules i.e. Mn, Fe, Co and CuPc and their interaction with (metallic) substrates and nonmagnetic TMPCs such as NiPc and ZnPc are only briefly mentioned.

6.1 Metal Substrates

One important characteristics of phthalocyanine and related molecules is the possibility to tailor the magnetic and electronic properties [1–6]. In view of the rapid development of electronics and the need for cheap and renewable applications molecule-substrate hybrid systems seem to be an attractive alternative to conventional electronics and are viewed as possible building blocks for future electronic and magnetic devices [7–10]. For many applications the orientation of the molecule on the substrate is of importance, e.g. standing upright or lying flat on the substrate. This influences coupling behavior, magnetism etc. Phthalocyanine molecules are good candidates because—under favourable conditions—they absorb flat on different types of substrates [8, 11–13]. CuPc and MnPc molecules are discussed for optical applications e.g. solar cells or transistors [14, 15] whereas Fe, Co and Cu phthalocyanine molecules are interesting for spintronic applications [2, 4, 8]. SnPc and VOPc molecules are exceptional cases because they are not flat as the other TM phthalocyanines, i.e. they have a lower symmetry than D_{4h} or D_{2h}, but also discussed in view of electronic devices [13, 16]. In this section we focus on the interaction of TMPc molecules with metal substrates discussing the adsorption of the molecules on different inert or reactive substrates. Depending on the substrate material, orientation, and reconstruction the electronic and magnetic properties of the adsorbed

H. C. Herper et al., *Molecular Nanomagnets*, Nanoscience
and Nanotechnology, https://doi.org/10.1007/978-981-15-3719-6_6

molecules can drastically differ from the properties of the free molecules discussed in Chap. 4. Since for application magnetic properties play an important role we try to shed light on the complex magnetic interactions and coupling between molecule and substrate.

6.1.1   Structural Aspects

In order to use Pc-substrate hybrid structures in applications often a flat orientation of the molecule on the substrate is desired. Near edge X-ray absorption fine structure (NEXAFS) measurements of the N K-edge and DFT calculations have shown that this is mostly fulfilled on metal substrates [8, 17–20]. However, in some cases out-of-plane orientation has been found, e.g. for FePc on Cu(001) or CuPc on a gold foil [17, 21]. There are basically two factors which determine the orientation of the molecule: the structure of the substrate and the growth rate. On (single) crystalline substrates the molecules tend to adsorb parallel to the surface whereas the use of polycrystalline (e.g. In-Sn-O) or amorphous (Au foil, oxidized Si) substrates lead to an out-of-plane orientation of the molecules [21]. Even on weakly interacting substrates planar deposition has been reported e.g. on HOPG (graphite) [22]. Obviously this growth mechanism is not related to the metal center of the molecule. It can be observed for molecules with nonmagnetic center such as SnPc [23] or even for metal free Pc (H_2Pc) [24]. Hence, the organic framework is responsible for the interaction with the substrate and the orientation parallel to the substrate [25]. However, with increasing coverage the molecular axis rotates out of the substrate plane, e.g. CuPc and H_2Pc on MoS_2 where with increasing thickness of the molecular film the molecules are tilted by 10° relative to the substrate [26] or in case of van der Waals bonded surfaces such as HOPG [27] on which thicker NiPc films show an α stacking as in Pc crystals, see Fig. 6.1a. Hence, a flat adsorption geometry as depicted on Fig. 6.1b will occur only as long as the molecule substrate interaction is stronger than the molecule–molecule interaction [28]. Therefore, the growth rate plays an important role for the orientation of the molecules. In case of higher (>0.5 ML/min) rates Pc molecules tend to cluster in the gas phase. Due to intramolecular couplings the interaction of the cluster with the substrate is much smaller than that of a single molecule [17]. It should be noted that a strong interaction with the substrate per se is not a guarantor for planar deposition. In some cases the interaction with the substrate is asymmetric which leads to a canted adsorption with one or more organic rings strongly interacting with the substrate while the rest of the molecule turns away from the surface [29].

In order to understand the influence of the substrate on the magnetic properties of individual molecules we focus on systems with low sub- to monolayer coverage where molecules can be adsorbed flat on the surface.

In case of NM metallic substrates such as Cu or Au one would expect less strong interaction between molecules and surface since there is no magnetic coupling between the two constituents. However, it turned out to be true only for very few surfaces in combination with particular Pc molecules [38] or in case of adlay-

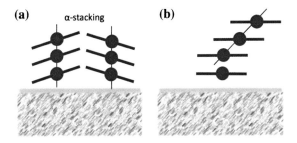

Fig. 6.1 Sketch of possible adsorption configurations on TMPc on a substrate. **a** Molecules stand upright on the substrate and layers grow in α-stacking known from bulk Pc (example NiPc/HOPG [27]) and **b** growth parallel to the substrate. At higher coverage molecules are shifted similar as in **a** but parallel to the substrate (for example CoPc/Pb(111), see Ref. [30]

ers on the metallic substrate, which increase the distance between the molecule and metal and therefore screen the interaction between them by reducing the hybridization between the out-of plane orbitals of molecule and substrate [8, 47]. It was observed from STM/IETS measurements [37] in agreement with DFT calculations [48] that an oxygen adlayer on a Cu(110) substrate reduces the interaction between the Cu substrate and adsorbed FePc molecules such that the molecules conserve basically their gas phase properties including the S = 1 spin state whereas the stronger interaction of the plain Cu(110) substrate destroys the magnetic moment of the molecule.

In general the Pc molecules are found to be chemisorbed on (transition) metal surfaces leading to a bigger overlap of electronic orbitals which causes changes in the geometrical, the electronic and the magnetic structure of the molecule. X-ray standing wave (XSW) experiments (for details of the method see for example Ref. [49]) have shown that MnPc is chemisorbed on Cu(001) ($d_{\mathrm{Mn-Cu}} = 2.24\,\text{Å}$) which agrees with the findings from van der Waals corrected DFT calculations [32]. However, based on DFT (PW91 [50]) calculations Zhang et al. claimed that all TMPc (TM = Mn, Fe, Co, Ni, Cu, Zn) are physisorbed on Au(111) with adsorption distances larger than 3.64 Å [41] which contradicts findings by Hu et al. obtained within the LDA [42]. These contradicting findings emphasize the importance of the choice of the exchange correlation functional and the van der Waals corrections within the theoretical description. Generalized gradient approximations such as PW91 [50] or PBE [51] lead to unphysical large adsorption distances whereas LDA tends to overbind the molecule, see Refs. [32, 41]. Taking into account van der Waals forces correct the adsorption distance [43] which is required for a proper description of possible charge transfer and the magnetic properties of the adsorbed molecule.

It turned out that in some cases the inclusion of van der Waals forces has not only influence on the adsorption distance but also on the adsorption site itself [47]. From DFT calculations for FePc on c2x2O/Co(001) the adsorption site was determined to be on top of the oxygen with molecule axes parallel to the surface axis (Fig. 6.2 structure A) if van der Waals corrections (D2 by Grimme [52]) are included. Without this additional term the molecule would prefer a position rotated by 30°–35° (cf.

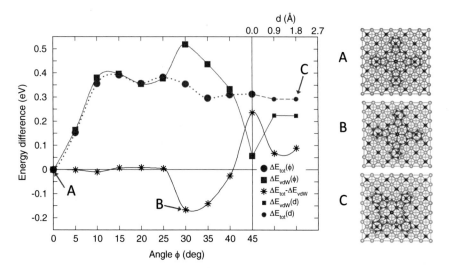

Fig. 6.2 Difference of the total energies ΔE if the FePc is rotated on the ground state position (left part) and moved from on top of O to the on top of Co position (45° rotated). Circles mark the differences in total energy if van der Waals forces are included and squares denote the plain van der Waals contribution to the energy. The difference between the two values is shown by the triangles and the minimum of this curve corresponds to the unphysical ground state configuration if van der Waals forces are neglected, for details see text. The structures on the right side depict the ground state on top of O position (A), the artificial minimum without van der Waals interactions (B), and the molecule on top of the Co which has the second lowest energy (C). Data taken from Ref. [47]

structure B in Fig. 6.2). This rotated structure would be much lower in energy than the unrotated configuration of the top of Co position (Fig. 6.2 structure C). At the same time the Fe moment would increase to $2.7\,\mu_B$ contradicting the experimental findings [47].

Transition metal surfaces offer a number of possible high symmetry adsorption positions for the Pc molecule, e.g. top of the metal, hollow site, bridge position, or in case of fcc(111) surfaces also hcp sites and the hybridization and exchange coupling between molecule and substrate will be affected by the adsorption site. Since the hybrid systems aim at minimizing the adsorption energy the adsorption site may differ even for very similar surfaces and metal centers. An overview of the adsorption positions of TMPcs on frequently used metal surfaces can be found in Table 6.1. Though, the distance and the angle between Pc molecule and substrate can be obtained, experimentally the determination of the adsorption site is more difficult. Only few XSW experiments exist and neither LEED (low energy electron diffraction) not STM experiments can give detailed information about the exact adsorption site of the molecule, since either the substrate is sharp or the molecules. However, the orientation of the molecule relative to the substrates axis can be obtained from STM experiments [19, 53] whereas DFT calculations are used to determine the adsorption position [35, 47]. Crosschecking of the experimental findings is sometimes done by

Table 6.1 Theoretically determined adsorption sites for transition metal phthalocyanine molecules TMPc on common metal substrates. The adsorption sites are denoted as fallows: hollow (H), top (T), bridge (B), hexagonal site (hcp). The angles give the relative deviation from the substrate axis. (DFT or STM+model), if not further specified the angle is named α

Surface		Transition metal center on TMPc					
		Mn	Fe	Co	Ni	Cu	Zn
Cu	(100)	B [31, 32]		T22.5° [33]		H/T26.5° [34]	
Cu	(110)		T 45° [35]				
Cu	(111)			B [36]		T [19]	
Cu	(100)/O[a]			T (O) [12]			
Cu	(110)/O[a]	30°(α) [37]					
Ag	(100)		H30° [38]	H30° [38]	H30° [38]	H30° [38]	
Ag	(110)		B30°[cc] [10]			H [39]	
			T45° [10]				
Au	(110)		T [40]	T [40]			
Au	(111)	Tα[41, 42]	Tα [41]	Tα [41]	Tα[41]	hcp [41, 42]	B [41]
			hcp [42]		hcp [42]		
Fe	(100)	T 45° [11]	T45° [11]			T45° [11]	
Fe	(110)		T [43]				
Co	(100)	B10° [44]	T45° [8]	(B)[b][45]		(B)[b] [45]	
		B [31]	B[b] [45]	(B)[b] [45]			
Co	(111)			B [36], Bα [46]		B [36]	
Co	(100)/O[a]		T (of O) [8, 47]				

[a] Surface reconstruction: $\sqrt{2} \times 2\sqrt{2}R45°$O/Cu(100); (2x1)O/Cu(110); c(2x2)O/Co(100)
[b] The bridge position proposed by Lach et al. contradicts to other results and the authors might not have taken into account all possible adsorption sites and orientations of the Fe-N$_1$ bond relative to the substrate
[c] Depending on the local coverage of the surface one of the two configurations is observed

comparing experimental UPS spectra to the ones obtained for the calculated ground state [54]. In rare cases such as for FePc/Au(111) the Kondo effect can be used to distinguish between different adsorption sites. For FePc molecules adsorbed on bridge or top positions the Kondo temperatures are quite different because due to the distinct arrangement of the substrate atoms under the TM atom the coupling between the impurity spin and the conduction electrons of the Au surface is different.

From the theoretical perspective the choice of the approximation for the exchange coupling functional as well as the effect of van der Waals forces might affect the results, despite that some general trends can be extracted. Transition metal Pcs deposited on Au(111) adsorb on top position with the Fe-N$_1$ bond rotated by an angle relative to the axis of the hexagonal lattice [41]. Exceptions are CuPc (hcp site) and ZnPc (bridge) which have a nearly full d shell. These findings seem to contradict the earlier results reported by Hu et al. [42], but they focussed only on high symmetry positions and found hcp and bridge positions for all TMPcs except MnPc (top). According to Ref. [41] these positions are close in energy to the ground state ($\Delta E = 10$–26 meV). Since Au(111) is a weakly reactive surface the molecules are physisorbed and the adsorption energies of different sites are quite small (<100 meV) independent whether LDA [42] or GGA(PW91) [41] was used. However, XSW experiments have shown that both methods fail to determine the adsorption distance between the CuPc molecule and the Au(111) substrate, i.e. within GGA the adsorption distance is about 20% too large whereas in case of LDA the adsorption distance amounts only to 3.04 Å being 7% too small. This discrepancy can only be overcome using a more sophisticated approach which covers the van der Waals character of the bonding. For CuPc on Au(111) Lüder et al. have shown that using the Tkatchenko–Scheffler method plus GGA in the formulation of PBE reduces the error in the adsorption distance to 1–2% [55]. An accurate description of the adsorption distance is not a self purpose but plays an important role in view of adsorption distance and site (Fig. 6.2) and has therefore significant influence on the electronic properties like charge transfer, changes of the magnetic moment, and possible structural deformation of the molecule and is even more important on more reactive surfaces such as Co(111) and Cu(111). On these substrates CoPc is chemisorbed on the bridge position [36, 56] with $d_{Co-surf}$ being 2.23 Å (2.70 Å) on Co(111) (Cu(111)) obtained from GGA plus Grimme's D2 approach for van der Waals corrections calculations [36]. From the broadening of the calculated PDOS and the blurry tunnelling densities from STM of CoPc/Cu(111) [57] it can be concluded that due to strong interaction with the substrate the former flat molecules are deformed by the substrate. Similar observations of a symmetry reduction due to molecular surface interaction have been made for CuPc on Cu(111) [19].

Though most DFT calculations focus on the interaction between an isolated molecule with the substrate material it should be mentioned that in experiment also molecule-molecule interaction comes into play, latest in the high submonolayer range (>0.6 ML) where molecules start forming dimers or small chains as discussed in Ref. [58] for CoPc on Au(111).

Using less stable substrate orientations such as Au(110) the substrate reconstruction changes with the molecule coverage [59, 60] and the Pc molecules tend to form self organized patterns on the surface, e.g. chains on Au(110) [40, 59, 61] or square like structures on the semiconducting InSb(100)-(4 × 4)c(8 × 2) surface [62] and on MoS$_2$ [26], see also Sect. 6.2. Theoretical calculations suggest an on top adsorption site for Fe and Co TMPcs on Au(110) and Cu(110) [35, 40] whereby the adsorption distance reaches from 2.6 Å(FePc/Cu(110) [35] to 3.2 Å(CuPc/Au(110)) [40]. Though these values are already significantly smaller than the ones obtained on the inert Au(111) surface they have to be viewed as upper boundaries since the GGA without further corrections has been used in the calculations i.e. the bonding is underestimated.

For the open (100) surfaces of fcc and bcc transition metal substrates different adsorption sites are reported. On Ag(100) TMPcs with TM = Fe, Co, Ni and Cu seem to prefer the hollow site position rotated by 30° relative to the axis of the substrate [20, 38]. This agrees with the observations from Lippel et al. by STM for CuPc on

Cu(100) [34] who observed a 26.5° rotation on the hollow or top position. The same adsorption site is observed for a bunch of different Pc molecules (TM, alkali Si center) adsorbed on Pt(100) and the authors of Ref. [63] concluded that the adsorption site is basically determined by the surface orientation and structure and less influenced by the central atom which is also in line with Ref. [25]. Obviously magnetism of the substrate plays also a role, because TMPcs adsorbed on ferromagnetic Fe and Co(100) substrates seem to prefer the top position [8, 11, 44, 47] with a strong overlap of the d-orbitals of both transition metal atoms and a direct magnetic coupling between the TMPc and the magnetic substrate. In all cases the TM-N_1 bond form a finite angle with the in-plane axis of the substrate. This theoretically determined rotations are in agreement with findings from STM measurements [33, 38]. What causes the rotation on the featureless (100) surface? Obviously the N_2 atoms (*aza*-N), cf. Fig. 1.1, which do not share a direct bond with the transition metal center of the molecule, play a crucial role [38]. The molecule tries to optimize the N_2-surface atom (here Ag) bond. On the magnetic Co(100) surface the situation is more complex compared to Ag(100) since magnetic coupling comes into play. DFT calculations (GGA+D2) by Herper et al. [47] have shown that not only the N_2 atoms hybridize with the substrate but also the C atoms of the outer benzene rings, see Fig. 6.3a. They show a strong coupling to the Co atoms whereas the contribution from the inner N_1 atoms is negligible. They are slightly pulled away from the surface. The crucial role of the outer rings regarding the orientation on the substrate becomes even more evident for FePc initially deposited on top position with Fe-N_1 bonds parallel to the axes, it relaxes to a local minimum 10° off the axes [47].

The adsorption sites summarized in Table 6.1 were almost all determined by DFT calculations with additional information from STM regarding the orientation relative to the substrate. Hence, strictly speaking they are valid at very low temperatures and for ideal substrates. Realistic surfaces possess terraces and step edges which seem to

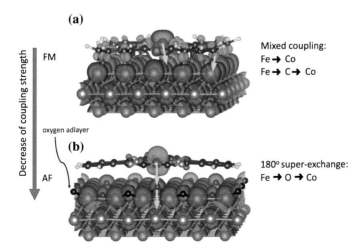

Fig. 6.3 Magnetic density of **a** FePc/Co(001) and **b** FePc/c(2×2)-O/Co(001). Data taken from Ref. [8]

be quite attractive for the Pc molecules [19]. A TMPc molecule which is located close to a step edge of the substrate try to place two of its 'legs' above the upper terrace [64], i.e. the molecule is bended and no longer flat on the substrate. Furthermore it should be mentioned that in experiment more than one adsorption position can occur on the one hand due to high coverages but on the other hand also because of locally varying coverages. For FePc on Ag(110) a combined STM/DFT study has shown that two adsorption positions exist (top and bridge) with different orientation relative to the substrate axis. Which configuration occurs where on the substrate is determined by the density of neighbouring molecules [10].

The above discussion of Pc adsorption of metal substrates suggests that in experiment not only the theoretically calculated adsorption site will be occupied. Especially at elevated temperatures the mobility of the molecules can be large enough to diffuse on the surface. This has especially been observed for adsorption on less reactive substrates like Au(111) and agrees with the theoretical findings of small energy differences between different adsorption sites [41, 42]. Thinking of molecule hybrid systems as building blocks for molecular electronics the diffusion is an unwelcome side effect but how to glue the molecules to one site and hinder diffusion? A possible solution could be the injection of electric pulses e.g. via an STM tip. Jeong et al. demonstrated STM experiments where the electric current from the tip dehydrogenates one or more of the organic C rings of a NiPc and the unsaturated carbon atoms form bonds with the Au atoms from the surface [65]. Similar observations have been reported by Stock et al. for CuPc on Cu(111) [66].

6.1.2 Magnetic Coupling

In case of magnetic substrates the molecules are chemisorbed and the bonding is largely determined by quantum mechanic exchange interactions between the metal ions of the molecule and the substrate. So far only few investigations of TMPc molecules on FM substrates exist but these studies demonstrate that direct coupling between the metal ions is the dominating mechanism [47, 67]. Even though the direct exchange through the metal ions dominates the coupling there is also a non negligible contribution from the rings. DFT calculations for FePc on Co(001) have shown that there is an indirect coupling contribution from the outer benzene rings of the Pc molecule which leads to a buckling of the molecule on the substrate [67], see Fig. 6.3a. Analogous to the observation on Ag(001) the *aza* N_2 atoms play an important role. They are almost located on top of a Co atom such that the benzene rings are centered around one Co atom of the surface. Despite the smaller AF coupling contributions from the rings the coupling between the metal ions is of FM nature [45, 67, 68]. The exchange coupling energies reported for FePc on Co(001) vary from 110 (bridge site) [45] to 293 meV (top of Co) [8] which might be related to the fact that the two papers discuss different adsorption sites for the molecules. Direct FM coupling between TMPc and Co(001) has also been observed for TM = Mn, Co, and Cu [31, 45].

In view of future molecule-based magnetic devices a controlled manipulation of the magnetic coupling between paramagnetic molecules such as TMPc and FM metal surfaces is desired. One possibility which has been discussed for TMPc and the slightly smaller TM porphyrin molecules is via surface adlayers [67, 69, 70]. Covering Co(001) with 0.5 ML of oxygen leeds to a c(2 × 2) reconstruction of the surface and the FePc molecule would be adsorbed on top on an O atom. The magnetic coupling turns into a 180° super exchange i.e. the partially filled d_π orbitals of Fe hybridize with the p orbitals of the O atom underneath and the O atom shares orbitals with the Co atoms in the (sub)surface layer, see Fig. 6.3b [8]. This leads to an AF coupling between the metal ions and has also been observed experimentally in XMCD experiments [8]. Due to the larger distance between the metal ions the coupling becomes also weaker. In case of FePc on c(2×2)-O/Co(001) from DFT (GGA+U, D2) calculations a reduction in coupling strength of more than 50% was obtained [8]. Similar results regarding the change in magnetic coupling due to an O adlayer have also been reported for other paramagnetic macrocycles [71, 72]. Hence the magnetic coupling is related to the distance between TMPc and the magnetic surface an accurate description of the adsorption distance is important. For example GGA (PBE) calculations for CoPc adsorbed on Fe(110) give zero spin for CoPc since the previously unoccupied d_{z^2} orbital hybridizes with the Fe d states. Including van der Waals corrections reduces the distance between molecule and substrate by 0.5 Å such that there is a bonding between the outer molecular rings and the substrate. This leads to a finite spin splitting of the molecule even though the Co center looses its moment which is in agreement with findings from SP-STM [43].

6.1.2.1 Electronic and Magnetic Properties

Due to hybridization effects and coupling between the molecule and the metallic substrate the electronic and magnetic structures of TMPc molecules can differ from the ones discussed in Chap. 4. Similar to the effect of ligands changes (see Chap. 7) in the spin state and charge transfer effects have to be expected whereby the situation is more complex than in the case of ligands. Due to the interaction with the substrate the out of plane d orbitals of the TM centers broaden compared to the gas phase density of states, see Fig. 6.4 and also the orbital occupation might change. In the example shown in Fig. 6.4a the highest occupied d state of FePc in gas phase and on Ag(100) (Fig. 6.4c) is a d_π whereas it has on the magnetic Co(001) substrate d_{xy} and d_z^2 character (Fig. 6.4b). Though the magnetic moment is in all three cases is about $2\,\mu_b$ the total magnetic moment of the molecule is quite different due to charge transfer effects on Ag(100), see below. In case of FePc on Ag(110) theory (DFT+GGA+U) predicts a S = 1 ground state with a total magnetic moment of $2\,\mu_B$ in contrast to the findings from XMCD which reveal $0.26\,\mu_B$ [74]. A similar reduction of the magnetic moment has been reported for FePc and CuPc on Au(110) [68]. As possible reason for this reduction a charge transfer between the 3d orbitals of the TM ion and the d_{z^2} orbital of the (110) oriented metal substrate is assumed [68,

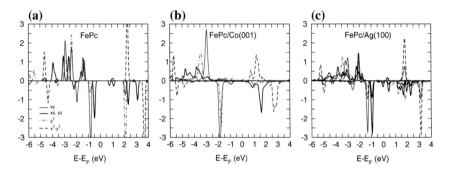

Fig. 6.4 Projected density of states of the Fe 3d states for the free FePc molecule (**a**) and the same molecule adsorbed on Co(001) (**b**) and Ag(001) (**c**). Calculations have been performed within the PBE and an effective U value of 3 eV on the Fe d orbitals. Data in **a** and **b** are taken from Refs. [73] and [47], respectively

74]. However, this effect is not captured in DFT+GGA calculations with Hubbard U corrections which give basically the same moment as in for the free molecule, cf. Chap. 4.

Even the interaction with quite inert surfaces such as Au(111) has strong influence on the magnetism of the molecule. From DFT calculations (LDA) and the observation of the Kondo effect it turned out that the magnetic moment of MnPc is hardly affected if adsorbed on Au(111), the magnetic moment in FePc is quenched to 50% of its gas phase value, and CoPc looses its moment completely [42, 57]. In case of CuPc DFT calculations show that the magnetic moment for the adsorbed molecule is the same as in gas phase, however no Kondo effect has been observed on Au(111) for CuPc. Li et al. concluded that this is because CuPc has only $d_{x^2-y^2}$ states near the Fermi level which do not interact with electrodes perpendicular to the surface [57].

The quenching of the magnetic moment in TMPcs adsorbed on substrates implies not necessarily that the magnetic moment of the transition metal center changes or vanishes. FePc on Ag(100) has a magnetic moment of 1.06 μ_B (GGA+U), caused by a large magnetic moment of the N and C atoms antiparallel aligned to the moment of the Fe center [38]. This means the substrate acts as charge reservoir for chemisorbed molecules where the distance between molecule and substrate is small and the p states of the ring atoms can hybridize with the substrate d states. It should be mentioned that the calculated charge transfer for all TMPc was about 1 electron but the influence on the magnetic moment of the TMPc molecule depends strongly on the occupation of the TM d states.

The above discussion of TMPcs on Au and Ag substrates demonstrates that the magnetic properties strongly depend on the choice of the surface material and orientation as well as on TM center. In order to understand this behavior the available calculated magnetic moments of the TMPcs on metal surfaces have been summarized in Fig. 6.5. The absolute values should be handled with care hence they depend at least partially on the respective method which has been used to calculate them, e.g.

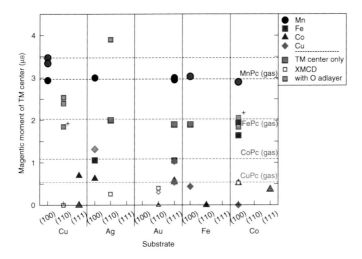

Fig. 6.5 Calculated magnetic moments of the TMPc adsorbed on different metal surfaces (filled symbols). Hatched symbols denote the magnetic moment of the TM center only and open symbols give the effective spin moments as obtained from XMCD measurements. Symbols with horizontal stripes denote magnetic moments obtained on surfaces with O-adlayers and additional + means only the moment of the TM ion is given. In some cases only the spin state was given which has been translated in a magnetic moment. The sources from which the data points have been extracted are summarized in Table 6.2 in the appendix

the type of potential or the exchange-correlation functional. However, overall chemical trends of the influence of the substrate layer on the magnetic of the molecules can be deduced.

The spin state of MnPc seems to be unaltered independent of the metallic substrate $S = 3/2$. The magnetic moment of the Mn ion varies from $2.90\,\mu_B$ on Co(001) and $3.47\,\mu_B$ on Cu(001) (both obtained from PBE) [31] whereby the large value on Cu(001) is due to the underestimated bonding in GGA. Newer calculations with PBE+U and van der Waals corrections (D2 [52]) for the same system give $3.34\,\mu_B$ on Cu(001) [32]. It is important to notice that GGA with van der Waals corrections alone would lead to an even stronger reduction of the moments because of missing screening effects. The actual moment of the adsorbed MnPc is slightly smaller due to spin polarisation of the organic rings, e.g. for MnPc on Cu(100) the N and C atoms carry a magnetic moment of $0.4\,\mu_B$ (PBE+U, D2) which is antiparallel aligned to the Mn moment [32]. Even though the spin state of Mn is not affected by the adsorption on simple metallic substrates, there is an obvious trend to reduce the moment on the Mn ion if the molecule is adsorbed on magnetic substrates, see Fig. 6.5. On NM substrates the magnetic moments is close to the value obtained in gas phase, whereas on FM substrates (Fe and Co) due to magnetic coupling between molecule and substrate the total moment is about 0.3–$0.6\,\mu_B$ reduced compared to the gas phase value.

In case of TMPc molecules with a more than half filled d shell the adsorption has a much larger influence on the magnetic properties. On Ag(100) and Au(111) FePc seems to reduce its magnetic moment by a factor of two because of a strong charge transfer to the organic rings and for (110) orientation this effect seems to be even larger, see Fig. 6.5. On magnetic substrates this effect is also reported but smaller e.g. for FePc at room temperature a Fe spin moment of $0.52\,\mu_B$ has been observed from XMCD. DFT calculations for FePc on Co(100) give total moment of $1.64\,\mu_B$ (PBE, U, D2) whereby the magnetic moment of the organic rings is $-0.37\,\mu_B$, but the magnetic moment of the Fe ion is basically the same as in the free molecule [47].

With increasing filling of the TM 3d shell the influence of the substrate on the magnetic properties of the adsorbed molecule becomes stronger. In most cases the magnetic moment is reduced compared to the gas phase value because Co and Cu have basically one unoccupied d orbital with out-of plane orientation (d_{z^2}) which is partially occupied due to hybridisation with the substrate orbitals. In some cases the moments are completely quenched, e.g. the moment of the Co ion in CoPc on Cu(111) (GGA+vdW) [36], see Fig. 6.5. Other than for MnPc the size of the magnetic moment seems not to be related to the magnetism of the substrate. The vanishing moment for CoPc/Fe(110) is an artefact of the method, i.e. missing van der Waals forces. It has been shown in Ref. [43] that the moment of the CoPc recovers if van der Waals corrections are included because the moment is located on the ring not on the TM center and only with Van der Waals forces the p-orbitals can overlap with the substrate orbitals. In nearly all other cases of TMPc with Co or Cu central atom the magnetic moment is at least reduced compared to the gas phase observations, cf. Fig. 6.5. The only exception was reported for CuPc on Ag(001). In this case the CuPc has a larger magnetic moment. As already mentioned for FePc this is a charge transfer effect. The consequences of the charge transfer depend on several factors: the filling of the TM d shell, the position of the lowest molecular orbital, and the hybridization of the molecular orbitals with the charge reservoir (substrate) [38].

So far only magnetic configurations of the plain molecules have been discussed, however, in view of applications a tailoring of spin states is desirable. For free molecules such as phthalocyanines (see Chap. 7) and porphyrins [75, 76] it is known that ligands have a strong impact on the spin state. Recently it has been shown that Li atoms added to TMPc/Ag(100) have a similar effect as ligands in gas phase. Due to reduction of the crystal field MnPc and FePc transform to the high spin state. An interesting effect occurs for NiPc and CuPc. NiPc is on Ag(001) in an S = 1 spin state whereas CuPc shown an S = 1/2 spin state. Doping the molecule substrate hybrid system with Li atoms the picture is reversed, i.e. Cu is nonmagnetic and NiPc is now in a S = 1/2 spin state [77].

An interesting effect occurs if the Ag(110) substrate is covered by an oxygen adlayer. Due to the O the magnetic moment of the FePc escalates from 2.0 (0.26) μ_B to 3.9 (2.1) μ_B in DFT calculations (XMCD). Even though the absolute theoretical values differ from the experimental findings the trend is the same. The Fe-O interaction provokes a filling of the majority spin channel and a depletion of the minority channel, i.e. the metallic Fe(II) character change to oxidized Fe(III) type [74]. This effect is reversible such that the systems switches to the high spin state if O_2 is added and back to S $=1$ be heating or adding H^+. As discussed in Sect. 6.1.2 on a magnetic substrate (Co(100)) the change is the spin state is not observed. Instead the magnetic coupling switches from strong FM to weak AF. If the magnetic coupling is reduced by adding an oxygen adlayer to the Co(100) substrate the charge transfer to the ring is reduced and the total magnetic moment of the FePc becomes 1.84 μ_B compared to 1.64 μ_B without O, see Table 6.2.

6.1.3 Spin Moments and the Spin Dipole Contribution

A comparison of the magnetic moments of the TM centers reported from theory with experimental values turned out to be quite difficult for adsorbed molecules. Most experimental data are obtained from XMCD measurements, see Sect. 3.6 from which one can extract only an effective spin moment m_s^{eff}. The effective spin moment differs from the actual spin moment by the spin dipole moment. The spin dipole operator T is defined by [78]

$$T = \sum_i Q^i m_s^i \tag{6.1}$$

where m_s^i is the spin moment and Q^i denotes the quadrupole tensor of the charge distribution for electron i. The later one is given by

$$Q^i = \delta - 3\hat{\mathbf{r}}_\alpha^i \hat{\mathbf{r}}_\beta^i \tag{6.2}$$

where α and β denote the spatial components. In case of bulk systems Q^i is of minor importance and $m_s \approx m_{\text{eff}}$. However, if the symmetry is reduced as for molecules [69] or clusters [79] adsorbed on a surface the spin dipolar contribution can become important [69, 80]. It is caused by the asperity of the spin density which is related to the crystal field in case of transition metals (where spin orbit coupling is weak). As long as the cubic symmetry is not broken Q vanishes. The experimentally reported spin moment can significantly differ from the theoretical observations, as for example for FePc on Co(001). Theoretical predictions for the Fe moment is about 1.64 to 1.94 μ_B [45, 47] whereas the effective spin moment obtained from XMCD measurements amounts to 0.52 μ_B (in-plane) [68].

In the following we focus only on the z component of the spin dipole operator. Its expectation value $\langle T_z \rangle$ is given by the trace of the density matrix multiplied by T_z. Assuming that for transition metals the spin orbit coupling is negligible, the size of $\langle T_z \rangle$ depends on the existence of a finite spin moment on the nonequivalent charge distribution on the orbitals.

In order to obtain the spin dipole moment and the effective spin moment van der Laan provided a scheme how to apply the general approach to a typical XMCD experiment such as in Fig. 6.6 [81]. It can be shown that the XMCD intensity depends on the relation between the magnetization direction $\mathbf{M}$, the polarization of the incident photon beam $\mathbf{P}$, and the surface normal $\mathbf{n}$. The angular dependence of the dipole operator is then given by

Table 6.2 Magnetic spin moments (m_s) of TMPc as obtained from DFT calculations, multiplet theory, and experiments. If not stated otherwise the values give the moment of the whole Pc molecule

Molecule	Surface	m_s (μ_B)	Method
MnPc	Free	3.47	DFT, PBE [31]
MnPc	Cu(100)	Free	DFT, PBE [31]
MnPc	Cu(100)	2.94	GGA, U, vdW [32]
MnPc	Au(111)	Free	VASP, PW91, [41]
MnPc	Au(111)	2.95	Dmol, LDA (VWN)[42]
FePc	Free	2.00	
FePc	Cu(110)	S = 0	STM, dI/dV curves, [88]
FePc	Ag(100)	1.06	DFT, PBE, U, vdW(D2), [38]
FePc	Au(110)	$0.4/n_h$	XMCD (magic angle), $B = 5$ T, single layer, [40]
FePc	Au(111)	1.90 (ion)	Kondo, STM, DFT PBE [89] (contradicting interpretation to [90])
FePc	Au(111)	1.05	Dmol, LDA (VWN) [42]
FePc	Au/Al$_2$O$_3$	0.64 (ion)	Film on Au surface, XMCD, B = 5T, T = 6K, [91]
FePc	Fe(100)	3.03	VASP, GGA+U, [11]
FePc	Co(100)	1.94	DFT, GGA, U [45]
FePc	Co(100)	1.64	DFT GGA, U, vdW (D2), [47]
FePc	Co(100)	0.52	Effective spin moment, T_z unknown [68]
FePc	O/Cu(110)	2.54	PBE, vdW-DF [48]
FePc	O/Cu(110)	2.40	PBE [48]
FePc	O/Cu(110)	S = 1	STM, dI/dV curves, [88]
FePc	O/Co(100)	1.84	DFT GGA, U, vdW (D2), [47]
CoPc	Free	1.00	[41]
CoPc	Cu(111)	0.70	PWSCF, [46] (most probably LDA)
CoPc	Cu(111)	0.00 (ion)	PWSCF, GGA+vdW, [36]
CoPc	Ag(100)	0.63	DFT, PBE, U, vdW(D2), [38]
CoPc	Au(110)	0.00	XMCD (magic angle), B = 5 T, single layer, [40]
CoPc	Au(111)	0.58	VASP, PW91, [41]
CoPc	Au(111)	0.54	Dmol, LDA (VWN)[42]
CoPc	Fe(110)	0.00	VASP, GGA [43]
CoPc	Fe(110)	Finite	VASP, GGA+vdW, [43]
CoPc	Co(100)	0.52	VASP, GGA, U [45]
CoPc	Co(111)	0.37 (ion)	PWSCF, GGA+vdW, [36]
CuPc	Free	0.54	[42]
CuPc	Ag(100)	1.32	DFT, PBE, U, vdW(D2), [38]
CuPc	Au(110)	$0.3/n_h$	XMCD (magic angle), B = 5 T, single layer, [40]
CuPc	Au(111)	Free	VASP, PW91, [41]
CuPc	Au(111)	1.03 (ion?)	Dmol, Becke exchange, LYP correlation [92]
CuPc	Au(111)	0.54	Dmol, LDA (VWN)[42]
CuPc	Fe(100)	0.44	VASP, GGA+U, [11]
CuPc	Co(100)	0.00	DFT, GGA, U [45]

Fig. 6.6 Sketch of typical setup of an XMCD experiment. The magnetization M and the polarization of the photon beam P define the incidence angle of the beam Φ. n stands for the direction of the surface normal

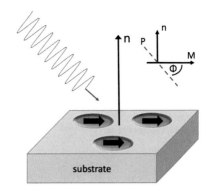

$$\langle 7T(\mathbf{n}, \mathbf{M}, \mathbf{P})\rangle = \frac{1}{4}\langle 7T_z\rangle\left[\cos(\phi) + 3\cos(\phi - 2\tau)\right] \tag{6.3}$$

where ϕ denotes the angle between the polarization $\mathbf{P}$ and the magnetization $\mathbf{M}$. The orientation between the magnetization and the surface normal $\mathbf{n}$ is given be the angle τ. For nonmagnetic substrate the alignment of the magnetic moments of the transition metal cents of the molecules is achieved by applying an external field which is usually parallel aligned to the polarization of the incident photon beam $\mathbf{M}\|\mathbf{P}$. In this special case Eq. 6.3 reduces to

$$\langle 7T(\tau)\rangle = \frac{1}{4}\langle 7T_z\rangle\left[3\cos^2(\phi) - 1\right] \tag{6.4}$$

and the experimentally observed effective spin moment reads

$$m_s^{\text{eff}} = m_s + \langle 7T(\tau)\rangle. \tag{6.5}$$

As can be seen from Fig. 6.7 T_z strongly varies with the incidence angle of the photon beam Φ. Only for $\Phi = 54.5°$, the so called magic angle, the argument in the brackets in Eq. 6.4 vanishes and the effective spin moment is identical with the actual spin moment, see Fig. 6.7. In the present example from a measurement at $\Phi = 70°$ (grazing incidence) would a spin moment of about 2.5 μ_B would been obtained in case of a fully saturated sample whereas the actual spin moment for FePc absorbed on Cu(001) is 1.92 μ_B.

It has been found that for molecules such as transition metal phthalocyanines [40, 80] or porphyrins [69] the dipole term can be of the same order of magnitude as the spin moment itself. For example, the spin moment of FePc on Ag/Al$_2$O$_3$ amounts to 0.65 μ_B (obtained at the magic angle) but for normal incidence ($\phi = 0$) $7\langle T_z\rangle = -0.52\,\mu_B$ such that the effective spin moment obtained from the sum rules is with 0.12 μ_B quite small. The same effect has been observed for CuPc on Ag(100) where the spin moment is about 0.5μ_B, but due to the spin dipole term XMCD measurements reveal at normal incidence provide an effective spin moment of about 1.67 μ_B [80].

Fig. 6.7 Calculated dipolar
contribution $7T(\Phi)$ and
effectice moment for FePc
adsorbed on Cu(001). The
dotted line marks the
calculated spin moment m_s.
The crossing point of m_s and
m_{eff} is the magic angle
where the dipolar
contribution vanishes

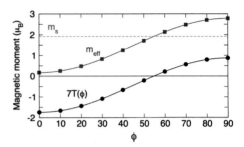

6.2 Nonmetallic and Complex Substrates

As discussed in the previous sections the electronic and magnetic properties of
adsorbed TMPc molecules strongly depend on the surface structure, orientation,
and reactivity. Similar observations have been made for more complex substrates.
Other than on metal substrates where with few exceptions the substrate controls the
structure of the molecule the situation is different on semiconducting substrates with
high surface reconstructions. CuPc deposited on Si(111)(7 × 7) destroys the (7 × 7)
superstructure of the Si surface [82]. Similar observations have been reported for
other semiconducting substrates. An overview can be found in Ref. [83]. For CuPc
or metal free H_2Pc adsorbed on MoS_2 it has been found that thin layers grow flat
on the graphite-like material but thick films tend to tilt by 10° [26]. Furthermore,
it has been observed from angular resolved UPS that the in-plane orientation does
not depend on the Pc center, i.e. H_2Pc and CuPc show the same growth pattern and
surface orientation. Both molecules show a square-like pattern rotated by 7°, 37°
or 67° relative to the crystal axis [26]. However, generally the growth pattern and
mechanism depend very well on the TM center of the Pc molecule. CuPc and oxo
vanadium Pc (VOPc) deposited on hydrogen terminated Si(111) with a slightly tilted
surface normal show very different growth behavior. The VOPc molecules grow in a
one-dimensional fashion along the steps of the surface whereas CuPc tends to form
clusters. A possible reason for this behavior could be the different size of the CuPc
which does not fit commensurably to the surface steps [84]. The fact that the problem of commensurability does not occur for VOPc might be related to the non flat
structure of the molecule. The O ligand sticks out of the molecular plane and forms
Si-O bond with the Si(111) surface [85]. In general semiconducting substrates tend
to form strong chemical bonds with the molecules, not necessarily via the TM center
but possible ligands as in the above discussed VOPc on Si(111) or via the N atoms
in the organic ring as observed for $CuPc/TiO_2(1 \times 2)$ [86].

Semiconducting substrates harbor also potential in view of spin switching systems if a strong molecule-surface coupling can be achieved. It has been predicted
theoretically in Ref. [13] that VOPc molecules adsorbed on Ga-rich GaAs(001) form
a strong bond between V and Ga via the O ligand of the molecule. This allows for
charge transfer and the spin state of the V atom changes from a doublet to triplet state
if adsorbed on the Ga-rich substrate. Mattioli et al. predicted that the spin state can

reversibly switched back to the doublet by hole injection in the substrate [13]. From the fact that no such spin switching could be observed for nonbonding VOPc, i.e. O pointing away from the substrate or flat Pc molecules such as CuPc it is obvious that the O ligand plays an important role, more precisely a charge accumulation at the O ligand occurs if the molecule is adsorbed on the GaAs substrate which leads then to the change in the magnetic moment and the spin state.

From applicational point of view recently TMPc-nanofiber heterostructures have attracted interest [3, 87]. FePc molecules with additional NO_2 end groups (TNFePc) have been deposited on TiO_2 nanofibers (plus H_2O_2) acts as a photocatalyst and is believed to be an efficient material for water purification from organic waste [87]. The same molecule has been proposed for oxidative desulfuration of diesel if combined with carbon nanofibers [3]. One significant advantage of the TMPc nanofiber heterostructures is the easy recycling which is quite important in view of the increasing permeation of technology.

Appendix

Details for the magnetic moments and spin state for transition metal Pcs on various substrates are shown in Table 6.2 including the theoretical approach or the measurement technique.

References

1. Sedona F, Marino MD, Forrer D, Vittadini A, Casarin M, Cossaro A, Floreano L, Verdini A, Sambi M (2012) Nat Mater 11:970. https://doi.org/10.1038/nmat3453
2. Heutz S, Mitra C, Wu W, Fisher A, Kerridge A, Stoneham M, Harker AH, Gardener J, Tseng HH, Jones T, Renner C, Aeppli G (2007) Adv Mater 19. https://doi.org/10.1002/adma.200701458
3. Xhen S, Lu W, Yao Y, Chen H, Chen W (2014) J React Kinet Mech Cat 111:535
4. Huang J, Xu K, Lei S, Su H, Yang S, Li Q, Yang J (2012) J Chem Phys 136:064707
5. Yang R, Gredig T, Colesniuc C, Schuller IK, Park J, Trogler BC, Kummel AC (2007) Appl Phys Lett 90:263506
6. Bohrer FI, Colesniuc CN, Park J, Ruidiaz ME, Schuller IK, Kummel A, Trogler WC (2009) J Am Chem Soc 13:478
7. Axtabal A, Ribeiro M, Parui S, Urreta L, Sagasta E, Sun X, Llopis R, Casanova F, Hueso LE (2016) Nat Commun 7:13751
8. Klar D, Klyatskaya S, Candini A, Krumme B, Kummer K, Ohresser P, Corradini V, de Renzi V, Biagi R, Joly L, Kappler JP, del Pennino U, Affronte M, Wende H, Ruben M (2013) Beilstein J Nanotechnol 4:320
9. Xiong ZH, Wu D, Vardeny ZV, Shi J (2004) Nature 427:821
10. Cai YL, Rehman RA, Ke W, Zhang HJ, He P, Bao S (2013) Chem Phys Lett 582:90
11. Sun X, Wang B, Yamauchi Y (2012) J Phys Chem C 116:18752
12. Guo Q, Huag M, Qin Z, Cao G (2012) Ultramicroscopy 118:17
13. Mattioli G, Filippone F, Bonapasta AA (2006) J Phys Chem Lett 1:2757
14. Brumboiu I, Totani R, de Simone M, Coreno M, Grazioli C, Lozzi L, Herper HC, Sanyal B, Eriksson O, Puglia C, Brena B (2014) J Phys Chem A 118:927

15. Dimitrakopoulos C, Mascaro DJ (2001) IBM J Res Dev 45:11
16. Lackinger M, Hietschold M (2002) Surf Sci Lett 520:L619
17. Zhang H, Huiqin Q, Dan J, Rehman AU, Dou W, Li H, He P, Bao S (2011) Chem Phys Lett 503:53
18. Petraki F, Peisert H, Aygül U, Latteyer F, Uihlein J, Vollmer A, Chassé T (2012) J Phys Chem C 116:11110
19. Karacuban H, Lange M, Schaffert J, Weingart O, Wagner T, Möller R (2009) Surf Sci 603:L39
20. Mugarza A, Lorente N, Ordejón P, Krull C, Stepanow S, Bocquet ML, Fraxedas J, Ceballos G, Gambardella P (2010) Phys Rev Lett 105:115702. https://doi.org/10.1103/PhysRevLett.105.115702
21. Peisert H, Schwieger T, Auerhammer JM, Knupfer M, Golden MS, Fink J, Bressler PR, Mast M (2001) J Appl Phys 90:466
22. Isvoranu C, Åhlund J, Wang B, Ataman E, Mårtensson N, Puglia C, Andersen JN, Bocquet ML, Schnadt J (2009) J Chem Phys 131:214709
23. Walzer K, Hietschold M (2001) Surf Sci 471:1
24. Nilson K, Ahlund J, Brena B (2007) J Chem Phys 108:7839
25. Wang B (2017) Mol Simul 43:384
26. Okudaira KK, Hasegawa S, Ishii H, Seki K, Harada Y, Ueno N (1999) J Appl Phys 99:6453
27. Ottaviano L, Nardo SD, Lozzi L, Passacantando M, Picozzi P, Santucci S (1997) Surf Sci 373:318
28. Molodtsova OV, Knupfer M, Ossipyan YA, Aristov VY (2008) J Appl Phys 104:083704
29. Altenburg SJ, Lattelais M, Wang B, Bocquet ML, Berndt R (2015) J Am Chem Soc 137:9452
30. Chen X, Fu YS, Ji SH, Zhang T, Cheng P, Ma XC, Zou XL, Duan WH, Jia JF, Xue QK (2008) PRL 101:197208
31. Javaid S, Bowen M, Boukari S, Joly L, Beaufrand JB, Chen X, Dappe YJ, Scheurer F, Kappler JP, Arabski J, Wulfhekel W, Alouani M, Beaurepaire E (2010) Phys Rev Lett 105:077201
32. Javaid S, Lebègue S, Detlefs B, Ibrahim F, Djeghloul F, Bowen M, Boukari S, Miyamachi T, Arabski J, Spor D, Zegenhagen J, Wulfhekel W, Weber W, Beaurepaire E, Alouani M (2013) Phys Rev B 87:155418
33. Qin ZH (2013) Chin Phys B 22:098108
34. Lippel PH, Wilson RJ, Miller MD, Wöll C, Chiang S (1989) Phys Rev Lett 62:171
35. Hu F, Mao H, Zhang H, Wu K, Cai Y, He P (2014) J Chem Phys 140:094704
36. Chen X, Alouani M (2010) Phys Rev B 82:094443
37. Tsukahara Y, Noto K, Ohara M, Shiraki S, Takagi N, Takata Y, Miyawaki J, Taguchi M, Chainani A, Shin S, Kawai M (2009) Phys Rev Lett 102:167203
38. Mugarza A, Robles R, Krull C, Korytár R, Lorente N, Gambardella P (2012) Phys Rev B 85:155437
39. Wei-Dong D, Fei S, Han H, Shi-Ning B, Qiao C (2008) Acta Phys Sin 57:628
40. Gargiani P, Rossi G, Biagi R, Corradini V, Pedio M, Fortuna S, Calzolari A, Fabris S, Cezar JC, Brookes NB, Betti M (2013) Phys Rev B 87:165407
41. Zhang Y, Du SX, Gao HJ (2011) Phys Rev B 84:125446
42. Hu Z, Li B, Zhao A, Yang J, Hou JG (2008) J Phys Chem C 112:13650
43. Brede J, Atodiresei N, Kuck S, Lazić P, Caciuc V, Morikawa Y, Hoffmann G, Blügel S, Wiesendanger R (2010) Phys Rev Lett 105:047204
44. Djeghloul F, Ibrahim F, Cantoni M, Bowen M, Joly L, Boukari S, Ohresser P, Bertran F, Fèvre PL, Thakur P, Scheurer F, Miyamachi T, Mattana R, Seneor P, Jaafar A, Rinaldi C, Javaid S, Arabski J, Kappler JP, Wulfhekel W, Brookes NB, Bertacco R, Taleb-Ibrahimi A, Alouani M, Beaurepaire E, Weber W (2013) Sci Rep 3:1272
45. Lach S, Altenhof A, Tarafder K, Schmitt F, Ali ME, Vogel M, Sauther J, Oppeneer PM, Ziegler C (2012) Adv Funct Mater 22:989
46. Iacovita C, Rastei MV, Heinrich BW, Brumme T, Kortus J, Limot L, Bucher JP (2008) Phys Rev Lett 101:116602
47. Herper HC, Bhandary S, Eriksson O, Sanyal B, Brena B (2014) Phys Rev B 89:085411. https://doi.org/10.1103/PhysRevB.89.085411

48. Hu J, Wu R (2013) Phys Rev Lett 110:097202
49. Zegenhagen J, Kazimirov A (eds) (2013) The x-ray standing wave technique, vol 7. Series on synchrotron radiation techniques and applications. World Scientific, Singapore
50. Perdew JP, Wang Y (1992) Phys Rev B 46:6671
51. Perdew J, Burke S, Ernzerhof M (1996) Phys Rev Lett 77:3865
52. Grimme S (2006) J Comput Chem 27:1787
53. Upward MD, Beton PH, Moriaty P (1999) Surf Sci 441:21
54. Huang YL, Wruss E, Egger DA, Kera S, Ueno N, Saidi WA, Bucko T, Wee ATS, Zojer E (2014) Molecules 19:2969
55. Lüder J, Eriksson O, Sanyal B, Brena B (2014) J Chem Phys 140:124711
56. Heinrich BW, Oacovita C, Brumme T, Choi DJ, Limot L, Rastei MV, Hofer WA, Kortus J, Bucher J (2010) J Phys Chem Lett 1:1517
57. Li W, Yu A, Higgins D, Llanos B, Chen Z (2010) J Am Chem Soc 132:17056
58. Cheng ZH, Gao L, ZT Deng, Jiang N, Liu Q, Shi DX, Du SX, Guo HM, Gao HJ (2007) J Phys Chem C 111:9242
59. Floreano L, Cossaro A, Gotter R, Verdini A, Bavdek G, Evangelista F, Ruocco A, Margante A, Cvetko D (2008) J Phys Chem C 112:10794
60. Fortuna S, Gargiani P, Betti MG, Mariani C, Calzolari A, Modesti S, Fabris S (2012) J Phys Chem C 116:6251
61. Betti MG, Gargiani P, Mariani C, Turchini S, Zema N, Fortuna S, Calzolari A, Fabris S (2012) J Phys Chem C 116:8657
62. Cox J, Bayliss S, Jones T (1999) Surf Sci 433–435:152
63. Cai Y, Qiao X (2014) Surf Sci 630:202
64. Li Z, Li B, Yang J, Hou JG (2010) Acc Chem Res 43:954
65. Jeong YC, Song SY, Kim Y, Oh Y, Kang J, Seo J (2015) J Phys Chem C 119:27721
66. Stock T, Nogami J (2014) Appl Phys Lett 104:071601
67. Klar D, Brena B, Herper HC, Bhandary S, Weis C, Krumme B, Schmitz-Antoniak C, Sanyal B, Eriksson O, Wende H (2013) Phys Rev B 88:224424
68. Annese E, Casolari F, Fujii J, Rossi G (2013) Phys Rev B 87:054420
69. Herper HC, Bernien M, Bhandary S, Hermanns CF, Krüger A, Miguel J, Weis C, Schmitz-Antoniak C, Krumme B, Bovenschen D, Tieg C, Sanyal B, Weschke E, Czekelius C, Kuch W, Wende H, Eriksson O (2013) Phys Rev B 87:174425. https://doi.org/10.1103/PhysRevB.87.174425
70. Miguel J, Hermanns CF, Bernien M, Krüger A, Kuch W (2011) J Phys Chem Lett 2:1455
71. Bernien M, Miguel J, Weis C, Ali ME, Kurde J, Krumme B, Panchmatia PM, Sanyal B, Piantek M, Srivastava P, Baberschke K, Oppeneer PM, Eriksson O, Kuch W, Wende H (2009) Phys Rev Lett 102:047202
72. Bhandary S, Brena B, Panchmatia PM, Brumboiu I, Bernien M, Krumme B, Etz C, Kuch W, Wende H, Eriksson O, Sanyal B (2013) Phys. Rev. B 88:024401
73. Brumboiu IE, Haldar S, Lüder J, Eriksson O, Herper HC, Brena B, Sanyal B (2015) J Chem Theory Comput
74. Bartolomé J, Bartolomé F, Brookes NB, Sedona F, Basagni A, Forrer D, Sambi M (2015) J Phys Chem C 119:12488
75. Brena B, Herper HC (2015) J Appl Phys 117:17B318
76. Liao MS, Huang MJ, Watts JD (2010) J Phys Chem A 114:9554
77. Stepanow S, Rizzini AL, Krull C, Kavich J, Cezar JC, Yakhou-Harris F, Sheverdyaev PM, Moras P, Carbone C, Ceballos G, Mugarza A, Gambardella P (2014) J Am Chem Soc 136:5451
78. Oguchi T, Shishidou T (2004) Phys Rev B 70(2):024412. https://doi.org/10.1103/PhysRevB.70.024412
79. Šipr O, Minár J, Ebert H (2009) EPL 87:67007
80. Stepanow S, Mugarza A, Ceballos G, Moras P, Cezar JC, Carbone C, Gambardella P (2010) Phys Rev B 82(1):014405. https://doi.org/10.1103/PhysRevB.82.014405
81. van der Laan G (1998) Phys Rev B 57:5250

82. Rochet F, Dufour G, Roulet H, Motta N, Scarlatta A, Piancastelli M, Crescenzi MD (1994) Surf Sci 319:10
83. Papageorgiou N, Salomon E, Angot T, Giovanelli JMLL, Lay GL (2004) Prog Surf Sci 77:139,170
84. Shimada T, Suzuki A, Sakurada T, Koma A (1996) Appl Phys Lett 68:2502
85. Eguchi K, Takagi Y, Nakagawa T, Yokoyama T (2013) J Phys Chem C 117:22843
86. Wang Y, Wu K, Kröger J, Berndt R (2012) AIP Adv 2:041402
87. Guo Z, Chen B, Mu J, Zhang M, Zhang P, Zhang Z, Wang J, Zhang X, Sun Y, Shao C, Liu Y (2012) J Hazard Mater 219–220:156
88. Tsukahara N, Noto K, Ohara M, Shiraki S, Takagi N, Takata Y, Miyawaki J, Taguchi M, Chainani A, Shin S, Kawai M (2009) PRL 102:167203
89. Gao L, Ji W, Hu YB, Cheng ZH, Dengand ZT, Liu Q, Jiang N, Lin X, Guo W, Du SX, Hofer WA, Xie XC, Gao HJ (2007) PRL 99:106402
90. Stepanow S, Miedema PS, Mugarza A, Ceballos G, Moras P, Cezar JC, Carbone C, de Groot FMF, Gambardella P (2011) Phys Rev B 83:220401(R)
91. Bartolomé LMGJ, Bartolomé F, Filoti G, Gredig T, Colesniuc CN, Schuller IK, Cezar JC (2010) PRB 81:195405
92. Zhao A, Li Q, Chen L, Xiang H, Wang W, Pan S, Wang B, Xiao X, Yang J, Hou JG, Zhu O (2005) Science 309:1542

Chapter 7
Influence of Ligands

Abstract The adsorption of small molecules on the metal center of 3d metal phthalo-cyanines alters the bonding scheme of the molecule and can induce different spin states and magnetic moments, as well as trigger more exotic effects like Kondo resonance. Soft X-ray spectroscopy studies and computational studies have highlighted this kind of effects in magnetic molecules like FePc, CoPc and MnPc by adsorption of for example CO, NO, O_2 molecules.

7.1 Phthalocyanine Molecules with Ligands

The adsorption of a ligand can substantially alter the electronic structure of open shell organo-metallic molecules like TMPc's (transition metal phthalocyanines), opening up for novel possible applications. The most reactive part of the molecule is the unsaturated TM in the center, which can be strongly influenced by chemical bonding. Axial ligands, i.e. molecules or adatoms adsorbed on the metal, can trigger a variation in the spin state of the metal ion, as well as the easy axis magnetization of the molecule, and in this way influence the magnetic properties of the molecule. In general, the adsorption of a small molecule on top of the metal ion generates a new bonding scheme, altering the D_{4h} square ligand field surrounding the metal. As a consequence the metallic 3d orbitals, involved in the new bonds, will be realigned in energy and will be differently populated.

The studies performed so far have been investigating the nature and the strength of the interaction and, in particular, soft X-ray spectroscopy and DFT calculations have addressed the changes in the TMPc's electronic structure in relation to the metallic states of the TMPc near the E_F. A summary of the recent studies is given below.

Tran and Kummer [1] studied the adsorption of NO and NH_3 on FePc by means of DFT calculations. The ligands resulted to be chemisorbed in both cases, although NH_3 only weakly. Charge transfer was revealed between Fe and the ligand, in particularly from Fe to NO and, conversely, from NH_3 to Fe. The bonding mechanism of NO on TMPc's was analysed by Nguyen et al. [2], who performed a DFT/GGA study of the adsorption of NO on isolated Mn-, Fe- and CoPc. The NO molecule results in all cases to be strongly chemisorbed, and to induce distortions in the molecular geom-

H. C. Herper et al., *Molecular Nanomagnets*, Nanoscience
and Nanotechnology, https://doi.org/10.1007/978-981-15-3719-6_7

etry, i.e. in a protrusion of the metal atom out of the molecular plane, also observed experimentally [3]. Among all TMPc's examined, the NO binds to the metal ion via the N atom, resulting in different orientations, which can be explained by the different partially filled 3d states close to E_F which participate in the bonding for each molecule. A study of the adsorption of CO, NO, O_2 and NH_3 on ML ordered films of a series of TMPcs including Mn-, Fe-, Co-, and CuPc adsorbed flatly on graphite was performed by Morishige et al. [4], concluding that all the TMPcs studied can in principle show a significant axial adsorptive activity with respect to these ligands considered.

A recent series of X-ray spectroscopy plus DFT/GGA studies by Isvoranu et al. have been dedicated to the adsorption of several small molecules on FePc deposited on Au(111) [5–7]. Based on spectroscopic results and DFT calculations. Isvoranu [5], it is argued that NH_3 attaches to FePc via the N atom that bonds to the Fe ion, involving the Fe d_{z^2} orbital and the N lone pair, resulting in a weak chemisorption, that causes a reshuffle in energy of the 3d electrons: the magnetic moment of the molecule passes according to GGA calculations from about 2.0 μ_B of the gas phase to the mere -0.3 μ_B with the ligand [5]. The adsorption of CO and NO on a ML of FePc on Au(111) was studied with the same techniques [6]. The bonding with the reactive CO and NO takes place with Fe via the N and C atoms. The rehybridization of the 3d metallic levels results in a reduction of the metal spin (a complete quenching in case of CO adsorption) and a weakening in the interaction with the substrate. Finally the effect of pyridine adsorption was analysed by the same group [7]. Pyridine also binds through the Fe ion and helps to manipulate the molecular magnetic moment, especially when achieved through a reversible and easily feasible mechanism, opening up avenues for novel applications in molecular spintronics, but not many works have explored this issue so far. The reversible control of the spin moment of individual MnPc by atomic H decoration has been reported in a recent STM study by Liu et al. [8]. The Kondo resonance could be switched on and off by adsorbing or dersorbing the H adatom through a local voltage pulse or thermal annealing, within a reversible process. The reversible change in the spin state of MnPc by means of CO adsorption has been reported by Strozecka et al. [9]. The study performed by combining STM and DFT calculations shows that the spin of MnPc through deposition on Bi(110) undergoes a reduction from S = 3/2 to S = 1, and a further reduction to S = 1/2 when CO is adsorbed. The original spin state is recovered via tip induced desorption of CO.

Although the study of O_2 adsorption on TMPc, and especially on FePc, is usually carried out to investigate catalytic properties, it is however a possible pathway to manipulate the molecular magnetic moment, as discussed for the other gases. Białek et al. [10] investigated the oxygen adsorption on FePc by means of ab initio calculations. A charge transfer from the metal to the O_2 is identified, and the metal-ion to oxygen bond involves Fe 3d and aromatic ring orbitals of e_g symmetry. An ellipsometry and DFT study of the electronic structure of FePc-O_2 by Friedrich et al. [11] shows a reordering of the orbital ordering of the 3d metal empty states in comparison to the single molecule in absence of the ligand. It is found that the triplet oxygen alters the spin of the compound since the O_2, couples antiferromagnetically to the MnPc leading to a $S = 1/2$ final spin. O_2 binding to Co- and FePc has been studied by Miedema et al. [12] by means of XAS, and XPS of the N 1s and metal 2p states, and Crystal Field Multiplet and time dependent DFT (TDDFT) calculations. The interaction with the oxygen molecule induces a change in the oxidation state in FePc, but not in CoPc, suggesting two different O_2 adsorption processes. CoPc is in a low spin state and configuration of its 3d states is unaffected by the O_2 adsorption, maintaining the ^2A type symmetry. FePc, on the contrary, switches from the intermediate spin to a high spin state with the O_2 bonding, the Fe^{2+} central ion evolving into a Fe^{3+}.

In all the cases when a new chemical bond is formed, the 3d electrons in the metal center hybridize with the orbitals of the small molecules. The distorted square ligand field as well as the resulting charge transfer between the ligand and the TMPc cause a reordering in the energy levels, and in the possibility of variations (usually partial quenching) in the total spin.

Another possible kind of doping that has been proposed is via Li atoms [13]. Monolayers of a series of TMPcs (MnPc, FePc, NiPc, CuPc) adsorbed on Ag(001) were doped by Li and studied by XMCD and atomic multiplet calculations. Li doping induces a reduction of the central ions from Cu(II), Ni(II), and Fe(II) ions to Cu(I), Ni(I), and Fe(I), except for Mn. This, combined with the change in the strength of the ligand field implies the possibility to switch on and off the magnetic moment of Cu and Ni. MnPc undergoes a change to high (5/2) spin, while in FePc ends up with a 10-fold increase of the orbital magnetic moment and a transition to a high-spin state [13].

References

1. Tran NL, Kummer AC (2007) J Chem Phys 127:2147011–2147017
2. Nguyen TQ, Escaño MCS, Kasai H (2010) J Phys Chem b 114:10017
3. Uchida K, Soma M, Onishi T, Tamaru KJ (1979) J Chem Soc Faraday Trans I 75:2839
4. Morishige K, Tomayasu S, Iwano G (1997) Langmuir 13:5184–5188
5. Isvoranu C, Knudsen J, Ataman E, Schulte K, Wang B, Boquet ML, Andersen JN, Schnadt J (2011) J Chem Phys 134:114711
6. Isvoranu C, Knudsen J, Ataman E, Schulte K, Wang B, Boquet ML, Andersen JN, Schnadt J (2011) J Phys Chem C 115:24718–24727
7. Isvoranu C, Wang B, Ataman E, Schulte K, Knudsen J, Andersen JN, Boquet ML, Schnadt J (2011) J Phys Chem C 115:20201–20208
8. Liu L, Yang K, Jiang Y, Song B, Xiao W, Li L, Zhou H, Wang Y, Du S, Ouyang M, Hofer WA, Neto AHC, Gao HJ (2013) Sci Rep 3:1
9. Stróżecka A, Soriano M, Pascual JI, Palacios J (2012) Phys Rev Lett 109:147202
10. Białek B, Brągiel (1996) Acta Phys Pol A 86:443–449
11. Friedrich R, Hahn T, Kortus J, Fronk M, Haidu F, Salvan G, Zahn DRT, Schlesinger M, Mehring M, Roth F, Mahns B, Knupfer M (2012) J Chem Phys 136:064704
12. Miedema PS, van Schooneveld MM, Bogerd R, Rocha TCR, Hävecker M, Knop-Gericke A, de Groot FMF (2011) J Phys Chem C 115:25422
13. Stepanow S, Rizzini AL, Krull C, Kavich J, Cezar JC, Yakhou-Harris F, Sheverdyaev PM, Moras P, Carbone C, Ceballos G, Mugarza A, Gambardella P (2014) J Am Chem Soc 136:5451

Chapter 8
Applications

Abstract In this chapter, a short introduction to possible applications utilizing organic molecules interacting with ferromagnetic substrates will be given. By explaining the spin filtering concept, studies with Cu-phthalocyanine as well as zinc methyl phenalenyl molecules will be reviewed. The hybridization of the electronic states of the molecules at the interface to those of substrates is a crucial point to realize functional hybrid metalorganic interfaces. A large magnetic anisotropy is found for the hybridized phenalenyl layer, which is important for the interface magnetoresistance effect used in this spin filter concept study.

8.1 Spin Filtering Concept

The concept of further miniaturisation of giant magnetoresistance (GMR) devices [1, 2] by utilising magnetic molecules is presented in Fig. 8.1 [3]. The molecular spin-valve model is presented in Fig. 8.1 (left). A magnetic source on the left produces a spin-polarized current. This flows through the magnetically oriented molecule and is then driven along a diamagnetic drain. The resistance of the entire system depends on the relative orientation of the magnetisation of the ferromagnetic source and the molecular magnet. This resistance is smaller for the case of parallel orientation as compared to an antiparallel orientation because of the spin-dependent scattering phenomena ($R_{\uparrow\uparrow} < R_{\uparrow\downarrow}$). A connected spin-valve concept is depicted in Fig. 8.1 (middle) [4]. On the top, the usual setup of a standard GMR system is given by two ferromagnetic layers being separated by a non-magnetic layer. The equivalent circuit diagram is shown in the middle for the spin-up current and the spin-down current as a parallel circuit. This explains the lower resistance for the parallel orientation of the magnetisations of the ferromagnetic layers. Such a device can also be realised by the connection of two single molecular magnets through a carbon nanotube as shown in Fig. 8.1 (middle) at the bottom. The characteristic feature of such a spin-valve is the 'butterfly' curve in the plot of the device current I versus the applied magnetic field Fig. 8.1 (right). The jump-like behaviour of the current corresponds to the switching of the magnetization of the individual magnetic layer of magnetic molecule that exhibits different switching fields H_1 and H_2.

H. C. Herper et al., *Molecular Nanomagnets*, Nanoscience
and Nanotechnology, https://doi.org/10.1007/978-981-15-3719-6_8

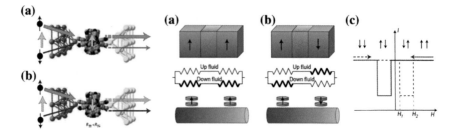

Fig. 8.1 Schematic presentation of the molecular spin filtering concept. Figures taken from Ref. [3, 4]

Historically at first, molecular multilayers were used in spin-valve devices before spin-valves with single molecular magnets were demonstrated. In the following section, we turn to the devices with multilayers of the molecular systems before we discuss the single molecule level.

8.2 Molecular Multilayers/Molecular Films

In 2004, Xiong et al. presented a spin-valve that contained two ferromagnetic electrodes (bottom electrode (FM_1): $La_{0.67}Sr_{0.33}MnO_3$ (LSMO) and top electrode (FM_2): Co), which were separated by a layer of a π-conjugated organic semiconductor (8-hydroxy-quinoline aluminium (Alq3)) [5]. This system is schematically presented in Fig. 8.2 (a) together with a scanning electron micrograph of the actual device (b). Indeed, the magneto-transport measurements show a 'butterfly' shape of the resistance versus the applied magnetic field. The relative change of the resistance of $\Delta R/R \sim 40\%$ is comparable to values achieved with metallic spin valves. Interestingly, Xiong et al. determined a lower resistance in the anti-parallel alignment and a larger resistance for the parallel alignment. This is opposite to the spin-valve effect, which is usually obtained when two identical FM electrodes are used. The authors assign this behaviour to the negative spin-polarization of the Co d-band [5]. From the detailed analysis of the magneto-transport measurements, a spin-diffusion length of $\lambda_S = 45$ nm is determined. This is one crucial property since a long spin-diffusion length is expected for organic semiconductors. The reason is the weak spin-orbit interaction and weak hyperfine interaction in these organic materials.

A different approach to achieve a more detailed insight into the spin injection process from a ferromagnet into an inorganic semiconductor is presented by Cinchetti et al. utilising 2-photon photoemission (2PPE) [6]. The basic principle is presented in Fig. 8.3 (left). Cu-phthalocyanine (CuPc) molecules with coverages ranging from sub-monolayers to multilayers are grown on Cobalt thin films on Cu(001). A first laser pulse with $h \cdot \nu = 3.1$ eV creates a spin-polarisation of the photoelectron in the Co film. However, because of the shortness of the electron mean free path by the

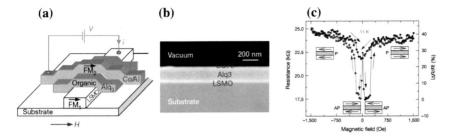

Fig. 8.2 **a** Schematic molecular spin filter with an organic interlayer (Alq3). **b** Scanning electron micrograph of one actual device (160-nm-thick Alq$_3$ spacer). **c** Magneto-transport measurement of the organic semiconductor spin-valve device. The magnetic field dependence exhibits the expected 'butterfly' shape for a spin valve at 11 K (130-nm-thick Alq3 spacer). Figures taken from Ref. [5]

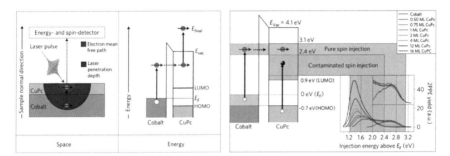

Fig. 8.3 Left: Schematic representation of the different length scales of the laser penetration depth as compared to the electron mean free path for the CuPc/Cobalt system together with a depiction of the 2-photon photoemission process (2PPE). Right: Scheme of the pure spin injection and contaminated spin injection regions linked to 2PPE spectra measured as a function of the CuPc thickness. Figures taken from [6]

second pulse, only the spin-polarised electrons at the surface-region in the CuPc are excited.

The actual 2PPE spin-integrated spectra are presented in Fig. 8.3 (right). Those present a pure spin-injection regime for the spin injection energies above E_F (E^i) larger than 2.4 eV. At smaller values of E^i, a 'contaminated' spin-injection regime is analysed. From the spin-polarised measurements, different values for the inelastic mean free path and the quasi-elastic spin-flip length were determined. By variation of the CuPc thickness, a quasi-elastic spin flip length, i.e. the mean distance one electron can travel before flipping its spin in a quasi-elastic scattering event, of $\lambda_{el,flip} = (12.6 \pm 3.4)$ nm could be determined when analysing the pure spin injection regime [6]. It is found by Cinchetti et al. that these quasi-elastic spin-flip processes are the dominant microscopic mechanisms limiting the spin diffusion length in CuPc.

Coming back to the idea of a spin valve utilising organic molecules, a fascinating concept is presented by Raman et al. [7]. The authors make use of zinc methyl phenalenyl (ZMP) molecules adsorbed on a Co films to create a functional hybrid metalorganic interface. When these molecules are grown on the ferromagnetic

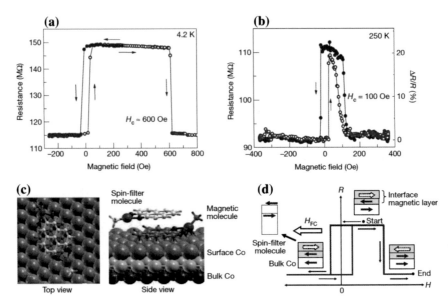

Fig. 8.4 Interface magnetoresistance effect as a function of the applied field for a Co (8 nm)/ZMP(35 nm)/Py (12 nm) device at 4.2 K (**a**) and 250 K (**b**). Ab initio model of the molecular hybrid system: top view (left) and side view (right) of the molecules (grey: carbon; red: oxygen; purple: zinc) on a Co(111) surface (**c**). Device model of the molecular dimer on the Co surface. The relative magnetization directions of the interface magnetic layer and the bottom bulk Co at various field values is presented (**d**). Figure taken from [7]

surface (Co film), a hybridized organometallic supramolecular magnetic layer develops. Interestingly, this hybridized layer exhibits a large magnetic anisotropy and spin-filter properties which are studied by magneto-transport measurements. Here, the interface magnetoresistance (IMR) effect is crucial. To study this effect, a Co (8 nm)/ZMP(35 nm)/Py (12 nm) device is analysed. The measurements are carried out at low temperature (4.2 K) as well as below room temperature (250 K) as presented in Fig. 8.4. The experimental results at 4.2 K (Fig. 8.4a) can be directly understood by a device model shown in Fig. 8.4d. The device consists of Co atoms that hybridize with a molecular dimer. This forms a magnetic molecule at the interface to the Co films whereas the upper molecule acts as a tunnel spin-filter molecule. To understand the hysteretic behaviour of the resistance (R) versus the applied magnetic field (H), the relative orientations of the magnetisation directions of the interface magnetic layer and the bottom bulk Co have to be inspected. By DFT calculations, it is found that in the first molecular layer, a net moment of $0.11 \mu_B$ is induced, which is oriented anti-parallel to the moment of the hybridized surface layer of Co. Hence, an anti-ferromagnetic (AFM) coupling is found for Co to this ZMP molecule. The magnetic field-dependence presented in Fig. 8.4a shows an independent switching of the interface magnetic layer indicating a large surface magnetic anisotropy. The interface hybridization is interpreted by Raman et al. as the source of the magnetic anisotropy energy making use of DFT results. It is concluded by the authors that the experimentally observed interface magnetoresistance effect is due to an enhancement of the magnetic anisotropy energy of the Co surface layer and a pronounced weakening of the magnetic exchange interaction between the surface Co layer and the layers below it [7]. This demonstrates the potential to tailor the magnetic exchange interaction at the interface on a molecular level.

References

1. Baibich MN, Broto JM, Fert A, Nguyen Van Dau F, Petroff F, Etienne P, Creuzet G, Friederich A (1988) J Chazelas Phys Rev Lett 61:2472
2. Binasch G, Grünberg P, Saurenbach F, Zinn W (1989) Phys Rev B 39:4828(R)
3. Bogani L, Wernsdorfer W (2008) Nat Mater 7:179
4. Sanvito S (2011) Nat Mater 10:484
5. Xiong ZH, Wu D, Vardeny ZV, Shi J (2004) Nature 427:821
6. Cinchetti M, Heimer K, Wüstenberg J-P, Andreyev O, Bauer M, Lach S, Ziegler C, Gao Y, Aeschlimann M (2009) Nat Mater 8:115
7. Raman KV, Kamerbeek AM, Mukherjee A, Atodiresei4 N, Sen TK, Lazic P, Caciuc V, Michel R, Stalke D, Mandal SK, Blügel S, Münzenberg M, Moodera JS (2013) Nature 493:509

Printed in the United States
By Bookmasters